D1088657

CHEMOTACTIC CYTOKINES

Biology of the Inflammatory Peptide Supergene Family

ADVANCES IN EXPERIMENTAL MEDICINE AND BIOLOGY

Editorial Board:

NATHAN BACK, *State University of New York at Buffalo*

IRUN R. COHEN, *The Weizmann Institute of Science*

DAVID KRITCHEVSKY, *Wistar Institute*

ABEL LAJTHA, *N.S. Kline Institute for Psychiatric Research*

RODOLFO PAOLETTI, *University of Milan*

Recent Volumes in this Series

A Continuation Order Plan is available for this series. A continuation order will bring delivery of each new volume immediately upon publication. Volumes are billed only upon actual shipment. For further information please contact the publisher.

CHEMOTACTIC CYTOKINES

Biology of the Inflammatory Peptide Supergene Family

Edited by

J. Westwick
The University of Bath
Avon, United Kingdom

I. J. D. Lindley
Sandoz Research Institute
Vienna, Austria

and

S. L. Kunkel
University of Michigan
Ann Arbor, Michigan

PLENUM PRESS • NEW YORK AND LONDON

Library of Congress Cataloging-in-Publication Data

International Symposium on Chemotactic Cytokines (2nd : 1990 : London, England)
Chemotactic cytokines : biology of the inflammatory peptide supergene family / edited by J. Westwick, I.J.D. Lindley, and S.L. Kunkel.
p. cm. -- (Advances in experimental medicine and biology ; v. 305)
"Proceedings of the Second International Symposium on Chemotactic Cytokines, held June 25-26, 1990, in London, United Kingdom"--T.p. verso.
Includes bibliographical references and index.
ISBN 0-306-44045-8
1. Cytokines--Congresses. 2. Chemotaxis--Congresses.
I. Westwick, J. II. Lindley, I. J. D. (Ivan J. D.) III. Kunkel, S. L. (Steven L.) IV. Title. V. Series.
[DNLM: 1. Chemotactic Factors--congresses. 2. Chemotaxis, Leukocyte--congresses. 3. Cytokines--congresses. 4. Inflammation--congresses. 5. Interleukin-8--congresses. W1 AD559 v. 305 / QW 690 I608c 1990]
QR185.8.C95I573 1990
616.07'9--dc20
DNLM/DLC
for Library of Congress 91-28671
CIP

Proceedings of the Second International Symposium on
Chemotactic Cytokines, held June 25-26, 1990,
in London, United Kingdom

ISBN 0-306-44045-8

© 1991 Plenum Press, New York
A Division of Plenum Publishing Corporation
233 Spring Street, New York, N.Y. 10013

All rights reserved

No part of this book may be reproduced, stored in a retrieval system, or transmitted in any form or by any means, electronic, mechanical, photocopying, microfilming, recording, or otherwise, without written permission from the Publisher

Printed in the United States of America

616.079
I61/8c

Major Sponsors

Parke-Davis
Pfizer
Sandoz
Upjohn Co.
Takeda Chemical Industries

Sponsors

Hoffman La Roche
ICI
Monsanto Co.
Miles Labs.
Proctor and Gamble Co.
Smith and Nephew
Synergen
Wellcome Research Labs.

Financial support

Abbot Labs.
Ciba-Geigy
Celltech
Fisons
Genentech
Immunex
Merck and Co.

FOREWORD

The existence of a new family of chemotactic cytokines was realised in 1987 following the isolation and structural determination by several groups of a peptide consisting of 72 amino acids which was a potent activator of neutrophils and a chemotactic agent for lymphocytes. The first symposium of this series was held at the Royal College of Surgeons of England in December 1988, entitled Novel Neutrophil Stimulating Peptides, and brought together the majority of the laboratories which had published in this area, see *Immunology Today* 10:146-147(1989).

Since the first symposium there has been a dramatic increase in our knowledge of the biology of this family of structurally related peptides.

The Second International Symposium on Chemotactic Cytokines was held at the Royal College of Surgeons of England in June 1990. The aim of this symposium was to provide both a forum for discussion and to determine whether this knowledge can be utilised in the design of novel therapeutic strategies for the treatment of inflammatory disorders. Although the majority of studies have been concerned with the regulation of these peptides at the molecular and cellular level, there is now evidence to suggest that specific members of this superfamily have a role in the pathogenesis of a number of diverse diseases including arthritis, psoriasis, atherosclerosis, wound repair, inflammatory lung diseases and glomerulonephritis.

We are very grateful to the companies listed on page v, without their support we would have been unable to organise the symposium and to present the proceedings in this book.

J. Westwick
I.J.D. Lindley
S.L. Kunkel

CONTENTS

ABSTRACTS

GRANULOCYTE AND MONOCYTE CHEMOTACTIC FACTORS: STIMULI AND PRODUCER CELLS

Jo Van Damme

Rega Institute
University of Leuven
B-3000 Leuven, Belgium

INTRODUCTION

The interest of our laboratory in cytokine production by various cell types originates from the early observation that different types of interferon (IFN) are released, depending on the inducer and cell type used. Indeed, infection of leukocytes with virus results in production of IFN-α (formerly leukocyte IFN), whereas fibroblasts release IFN-β (formerly fibroblast IFN) in response to double-stranded RNA (dsRNA) or virus (1). Further studies on IFN allowed us to discover a leukocyte-derived protein, that exerts an indirect antiviral activity through IFN-β induction on fibroblasts. The molecule involved could be identified as interleukin-1β (IL-1β) (2), now well known as an inducer of other cytokines such as IL-6 (3). In contrast to IFN, IL-6 exists as a single molecular species, but can be induced in a variety of cells, including fibroblasts and leukocytes. These cell types also produce granulocyte (GCP/IL-8) and monocyte (MCP) chemotactic proteins in response to IL-1, other cytokines and cytokine inducers (Table 1). In addition, different tumor cell lines are capable of releasing these chemotactic molecules.

MATERIALS AND METHODS

Induction of Chemotactic Activity

Human diploid fibroblasts (E_1SM, a strain of embryonic skin and muscle cells), the osteosarcoma cell line MG-63, the epidermal carcinoma cell line HEp-2, the hepatoma cell line Malavu and the melanoma cell line Bowes were grown in Eagle's MEM (EMEM) with Earle's salts, supplemented with 10% (v/v) FCS. For the induction experiments, confluent monolayers grown in 24-well (1.9 cm^2) dishes were incubated with EMEM (0.5 ml/well) containing 2% (v/v) serum. The cells were stimulated with different concentrations of the following inducers: natural human IL-1β, purified to homogeneity (2); the double-stranded RNA poly rI:rC (P-L Biochemicals, Inc., Milwaukee, WI); measles virus [Attenuvax strain, $10^{5.6}$ 50% tissue culture infectious dose/ml ($TCID_{50}$/ml)]; and *Escherichia coli* (strain NCIB 8473, 15 x 10^{10} cells/ml).

Peripheral blood mononuclear cells were isolated by gradient centrifugation (400 x g, 30 min) on Ficoll-sodium metrizoate (Lymphoprep, Nyegaard, Oslo, Norway).

Chemotactic Cytokines, Edited by J. Westwick *et al.*
Plenum Press, New York, 1991

Table 1. Cellular sources of GCP/IL-8 and MCP

Cell type	Inducers of	
	GCP/IL-8	MCP
Mononuclear leukocytes	LPS, ConA, PHA, PMA, IL-1, TNF (4-19)[1]	LPS, ConA, PHA, IL-1, IFN- (27-34)
Fibroblasts	virus, dsRNA, IL-1, TNF (12-14)	virus, dsRNA, PDGF, IL-1, TNF (27,35-37)
Endothelial cells	LPS, IL-1, TNF (15-18)	LPS, IL-1, TNF, EGF (37-39)
Chondrocytes	IL-1, TNF, poly rI:rC (19)	
Synovial cells	IL-1 (20)	
Keratinocytes	IL-1, IFN- + TNF (21,22)	
Epithelial cells	IL-1, TNF (23)	
Smooth muscle cells		- (40)
Leukemia	PMA, DMSO (24)	LPS (41)
Sarcoma	IL-1, virus (12)	virus (27,42,43)
Carcinoma	- (25)	- (42)
Hepatoma	IL-1, TNF, virus (26,27)	IL-1 (27)
Glioma		- (44)

[1]Literature references.

Abbreviations: LPS: lipopolysaccharide; ConA: Concanavalin A; PHA: phytohemagglutinin; PMA: phorbol 12-myristate 13-acetate; dsRNA: double-stranded RNA; polyrI:rC: polyriboinosinic:polyribocytidilic acid; DMSO: dimethyl sulfoxide; TNF: tumor necrosis factor; PDGF: platelet-derived growth factor; EGF: epidermal growth factor;

The adhering cell population was obtained by incubating the mononuclear cell fraction in stationary cultures for 2 h at 37°C. Non-adherent cells were then removed by repeated washes and the cultures were replenished with EMEM. The adherent cell fraction consisted of 80% monocytes as tested microscopically using PE-conjugated human Le-M3 antibody (Becton Dickinson, Mountain View, CA). Monocyte cultures were stimulated for 48 h at 37°C with IL-1β (2), lipopolysaccharide (LPS) from *E.coli* (Difco, Detroit, MI) or Concanavalin A (ConA, Calbiochem, San Diego, CA).

Assay for Chemotaxis

Polymorphonuclear neutrophils and mononuclear cells in heparinized human peripheral blood from a single donor were separated by gradient centrifugation (30 min, 400 x g) on Ficoll-sodium metrizoate. The total mononuclear cell fraction was used as a source for monocytes. The pellet, containing granulocytes and erythrocytes, was suspended in hydroxyethyl starch (Plasmasteril, Fresenius AG, Bad Homburg, FRG) for 30 min to remove erythrocytes by sedimentation. Residual erythrocytes were eliminated by lysis in bidistilled water (30 s). Purified neutrophils were finally obtained by centrifugation (30 min, 20,000 x g) in a Percoll gradient (d = 1.054 g/ml). Granulocytes and mononuclear cells were washed, counted and re-suspended at 2 x 10^7 and 1 x 10^8 cells/ml, respectively, in HBSS (Gibco, Paisley, Scotland), supplemented with pyrogen-free human plasma protein (1 mg/ml albumin, Cohn fraction V).

Chemotactic activity for monocytes and for granulocytes was measured under agarose according to the method of Nelson *et al.* (45). Briefly, mononuclear cells (10^6/well) and granulocytes (2 x 10^5/well) were exposed to serial dilutions of test samples (10 µl/well) and to control medium (HBSS plus albumin). Pure GCP/IL-8 (8) and MCP (27) were used as positive controls for granulocyte and monocyte chemotactic activity, respectively. After 2h incubation at 37°C, cells were fixed and the migration distance was scored microscopically. Effective migration represents the difference between the migration distance towards the test sample and the spontaneous migration distance towards the control medium. The titration end point, corresponding to 1 U/ml (± 10 ng/ml), was calculated from a dilution resulting in a half-maximal effective migration distance as compared to that obtained with GCP/IL-8 or MCP. In crude cell supernatants chemotactic activity was expressed as percentage of maximal cell migration obtained with GCP/IL-8 or MCP.

Separation of Granulocyte and Monocyte Chemotactic Activity

Chemotactic activity in crude supernatant from stimulated fibroblast cultures was first concentrated and partially purified by batch adsorption to controlled pore glass beads (CPG-10-350, Serva, Heidelberg, FRG). The eluates from controlled-pore glass were then applied to a heparin-Sepharose CL-6B (Pharmacia, Uppsala, Sweden) column from which the bound proteins were eluted by a linear NaCl gradient (0.05-1.0 M) followed by a step-wise gradient to 2 M NaCl. Peak fractions (10 ml in total) from the heparin-Sepharose column were dialyzed against 50 mM formate, pH 4.0 and then subjected to FPLC, using a Mono S cation-exchange column (Pharmacia). After an initial wash with 50 mM formate, pH 4.0, chemotactic activity was eluted (1 ml/min, 1 ml fractions) with a linear NaCl gradient (0.05-1.0 M) in the same buffer. Absorbance was monitored at 280 nm.

RESULTS AND DISCUSSION

Simultaneous Induction of GCP/IL-8 and MCP in Fibroblasts and Adherent Mononuclear Cells

Peripheral blood mononuclear cells stimulated with LPS, ConA or IL-1β produced chemotactic activities for both monocytes (MCP) and granulocytes (GCP) as measured by cell migration under agarose (Table 2) (8,45). Mononuclear cells were better producers of GCP than of MCP. The optimal dose for induction with ConA and IL-1 was 10 µg/ml and 50 U/ml, respectively, whereas for LPS all doses tested significantly induced GCP and MCP. None of these inducers had direct *in vitro* chemotactic activity. Purification of the activities revealed that the granulocyte and monocyte chemotactic factor did reside in two different but related molecules. GCP was identified by NH2-terminal sequence analysis as neutrophil activating factor-1/interleukin-8 (NAP-1/IL-8) from monocytes (4-11,46), whereas MCP was found to be identical to the chemotactic molecule isolated from tumor cells (28-30).

Diploid fibroblasts produced GCP and MCP in response to various stimuli, including IL-1β, the dsRNA polyrI:rC and measles virus (Table 3). The induction was dose-dependent, 10-100 U/ml of IL-1β, 100 µg of polyrI:rC and $10^{4.6}$ tissue culture infective dose 50% ($TCID_{50}$) of measles virus being the most effective

Table 2. Induction of GCP/IL−8 and MCP in adherent peripheral blood mononuclear cells

Inducer[a]	Dose	Chemotactic activity (%)[b]	
		GCP	MCP
IL−1β	50 U/ml	25	18
	5	25	< 10
	0.5	< 10	< 10
LPS	50 μg/ml	63	23
	5	96	42
	0.5	79	20
ConA	10 μg/ml	88	40
	1	50	32
	0.1	17	10

[a]Fresh mononuclear cells, isolated by centrifugation on Ficoll−sodium metrizoate and adherence to plastic, were stimulated for 48 h.
[b]Percentage of migration distance obtained by pure human GCP/IL−8 and MCP (n=3).

Table 3. Dose−dependent induction of GCP/IL−8 and MCP in human fibroblasts

Inducer[a]	Dose	Chemotactic activity (%)[b]	
		GCP	MCP
IL−1β	100 U/ml	45	66
	10	67	42
	1	21	17
	0.1	< 10	< 10
Poly rI:rC	100 μg/ml	65	75
	10	16	49
	1	12	26
Measles	$10^{4.6}$ $TCID_{50}$/ml	75	59
	$10^{3.6}$	33	42
	$10^{2.6}$	< 10	<10

[a]Confluent monolayers of diploid cells were induced for 48 h.
[b]Percentage of migration distance obtained by pure human GCP/IL−8 and MCP (n=3).

doses. The maximal induction capacities of these substances were equal and the relative amounts of GCP and MCP produced were comparable. MCP and GCP could be separated by cation exchange chromatography (Fig. 1). GCP from fibroblasts was shown to be serologically related to monocyte−derived NAP−1/IL−8. Furthermore, except for differences in NH_2−terminal processing, fibroblasts−

derived GCP has an identical amino acid sequence as NAP-1/IL-8 (12). Purification of MCP from diploid cells revealed an elution profile identical to that of MCP from mononuclear cell. Finally, MCP isolated from a fibroblastoid cell line was found to have the same amino acid sequence as MCP from peripheral blood leukocytes (27,30).

Thus, fibroblasts and adherent mononuclear cells can both generate GCP/IL-8 and MCP. There exists, however, specificity in that MCP and GCP are only chemotactic for monocytes or granulocytes, respectively. Specificity in production of GCP/IL-8 and MCP became obvious when the different inducers were compared on the two cell types. LPS or *E.coli* induced GCP/IL-8 and MCP in mononuclear cells but not in fibroblasts (27). On the contrary, polyrI:rC, measles virus and IL-1β are better inducers of GCP/IL-8 and MCP in fibroblasts than in mononuclear cells. This illustrates that the induction of chemotactic factors is also inducer-specific. With respect to the *in vivo* situation, it therefore seems that mononuclear cells would produce GCP/IL-8 and MCP after bacterial infection, whereas fibroblasts would rather generate chemotactic factors after viral infection. Since under similar conditions fibroblasts can also generate several other cytokines, including IFN-β, IL-6 and colony stimulating factors (47), these cells can contribute indirectly to the inflammatory response. The cytokine IL-6, inducible in fibroblasts and monocytes by cytokines and infectious agents is not an intermediate for GCP/IL-8 or MCP production since by itself it could not induce chemotactic activity in these cells (27).

The induction of MCP and GCP/IL-8 protein in human mononuclear cells and fibroblasts was confirmed by measuring mRNA levels for these chemotactic factors (13,21,36,37). However, a disparate gene expression of MCP and GCP/IL-8 was observed in mononuclear cells (48). LPS challenged monocytes did express mRNA for GCP/IL-8 but failed to express mRNA for MCP. In contrast, B-lymphocytes expressed mRNA for MCP but not for GCP/IL-8. This illustrates a cell specific production of MCP and GCP within mononuclear cell populations.

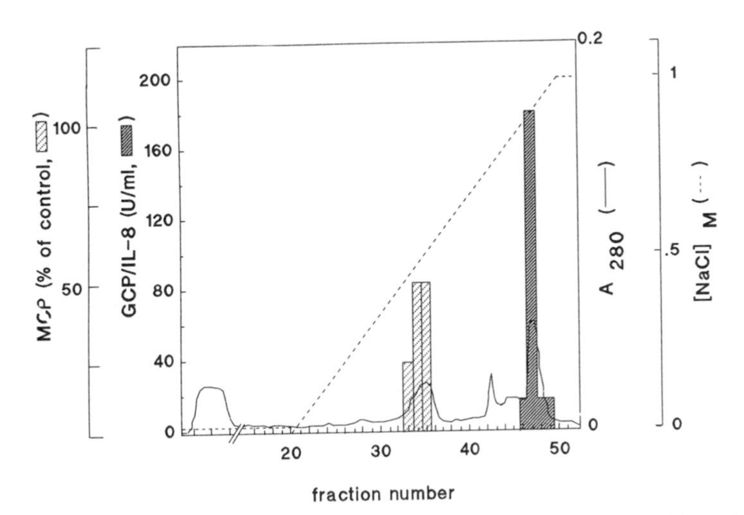

Figure 1. Separation of MCP and GCP/IL-8 from diploid fibroblasts by cation-exchange FPLC at pH 4.0. Chemotactic activities (histograms) were eluted in a linear NaCl gradient (---) and the absorbance was monitored at 280 nm (——).

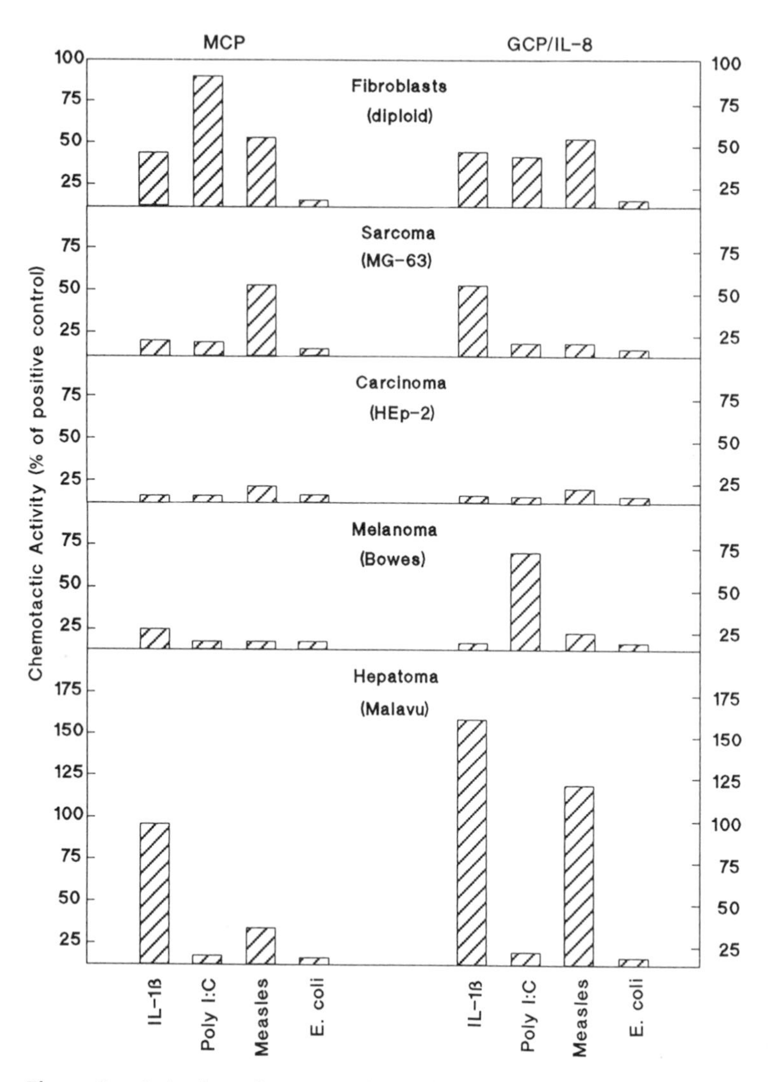

Figure 2. Induction of MCP and GCP/IL-8 in tumor cells. Confluent monolayers were stimulated with IL-1β (100 U/ml), poly rI:rC (100 μg/ml), measles virus ($10^{4.6}$ $TCID_{50}$/ml) or *E.coli* (10^9 cells/ml) for 48 h.

Production of GCP/IL-8 and MCP by other human cell types and tumor cell lines

In addition to mononuclear cells and fibroblasts GCP/IL-8 and MCP can also be produced by many other cell types in response to a variety of inducing substances, including LPS, dsRNA, virus or the cytokines IL-1 and TNF-α (Table 1). The induction of GCP/IL-8 in chondrocytes and synovial cells (19,20) by IL-1 and TNF-α points to a role of these cytokines in rheumatoid arthritis. The release of GCP/IL-8 from these cell types could in part explain the infiltration of leukocytes into the synovium. Secretion of GCP/IL-8 by dermal fibroblasts and keratinocytes (12-14,21) *in vitro* correlates with its presence during skin diseases. Intradermal injection of GCP/IL-8 is associated with a prompt infiltration of neutrophils into the skin (8,50). The production of GCP/IL-8 by endothelial cells (15-18) and its identify with leukocyte adhesion inhibitor (17) might be related to the neutrophilia induced after intravenous injection of GCP/IL-8 (8).

Infiltration of leukocytes into inflamed tissue or into tumors could be mediated through the release of GCP/IL-8 and MCP by surrounding tissue or tumor cells themselves. In neoplastic tissue a correlation between monocyte chemotactic activity and macrophage content has been made (50) and various tumor types have been shown to produce GCP/IL-8 and MCP (Table 1). Fig. 2 shows that GCP/IL-8 and MCP can be induced in tumor cell lines by cytokine, virus or dsRNA. In contrast to diploid fibroblasts that secreted GCP/IL-8 and MCP in response to all these substances, tumor cell lines, depending on their origin, produced often only one of the two activities in response to one particular inducer. For example, MG-63 osteosarcoma cells were stimulated to produce MCP after treatment with measles virus, whereas IL-1β was the better inducer of GCP/IL-8 in these cells. The Bowes melanoma cell line produced no MCP, but could be induced to release GCP/IL-8 by induction with poly rI:rC. Still another response was observed with Malavu hepatoma cells, that released both GCP/IL-8 and MCP after induction with IL-1β or measles virus. In contrast, HEp-2 carcinoma cells could not be stimulated to produce either GCP/IL-8 or MCP. *E.coli*, that can induce MCP and GCP/IL-8 in mononuclear cells, failed to do so in the cell lines tested. Thus, the production of chemotactic factors by tumor cells both depend on the type of tumor, inducer and released factor tested. In addition, others have demonstrated that different tumor cell types often spontaneously release monocyte chemotactic activity (42,43). Since IL-6 production by tumor cell lines is still differently regulated (27), this further indicates for a specificity in cytokine release by tumors.

Acknowledgements: The excellent editorial help of Christiane Callebaut, Dominique Brabants and the skilful technical assistance of Jean-Pierre Lenaerts and Rene Conings are gratefully acknowledged. The author thanks Dr. Ghislain Opdenakker for critical reading of the manuscript. J. Van Damme is research associate of the Belgian National Fund for Scientific Research (NFWO).

REFERENCES

1. Havell, E.A., B. Berman, C.A. Ogburn, K. Berg, K. Paucker and J. Vilcek. 1975. Two antigenically distinct species of human interferon, *Proc. Natl. Acad. Sci. USA* 72: 2185-2187.
2. Van Damme, J., M. De Ley, G. Opdenakker, A. Billiau, P. De Somer and J. Van Beeumen. 1985. Homogeneous interferon-inducing 22K factor is related to endogenous pyrogen and interleukin-1, *Nature* **314**: 266-268.
3. Van Damme, J., G. Opdenakker, R.J. Simpson, M.R. Rubira, S. Cayphas, A. Vink, A. Billiau and J. Van Snick. 1987. Identification of the human 26-kD protein, interferon β-2 (IFN-β2), as a B cell hybridoma/plasmacytoma growth factor induced by interleukin 1 and tumor necrosis factor, *J. Exp. Med.* **165**: 914-919.
4. Schmid, J. and C. Weissmann. 1987. Induction of mRNA for a serine protease and a β-thromboglobulin-like protein in mitogen-stimulated human leukocytes, *J. Immunol.* **139**: 250-256.
5. Yoshimura, T., K. Matsushima, S. Tanaka, E.A. Robinson, E. Appella, J.J. Oppenheim and E.J. Leonard. 1987. Purification of a human monocyte-derived neutrophil chemotactic factor that has peptide sequence similarity to other host defense cytokines, *Proc. Natl. Acad. Sci. USA* **84**: 9233-9237.
6. Schröder, J.-M., U. Mrowietz, E. Morita and E. Christophers. 1987. Purification and partial biochemical characterization of a human monocyte-derived, neutrophil-activating peptide that lacks interleukin 1 activity, *J. Immunol.* **139**: 3474-3483.
7. Peveri, P., A. Walz, B. Dewald and M. Baggiolini. 1988. A novel neutrophil-activating factor produced by human mononuclear phagocytes, *J. Exp. Med.* **167**: 1547-1559.
8. Van Damme, J., J. Van Beeumen, G. Opdenakker and A. Billiau. 1988. A novel, NH_2-terminal sequence-characterized human monokine possessing neutrophil chemotactic, skin-reactive, and granulocytosis-promoting activity. *J. Exp. Med.* **167**: 1364-1376.
9. Matsushima, K., K. Morishita, T. Yoshimura, S. Lavu, Y. Kobayashi, W. Lew, E. Appella, H.F. Kung, E.J. Leonard and J.J. Oppenheim. 1988. Molecular cloning of a human monocyte-derived neutrophil chemotactic factor (MDNCF) and the induction of MDNCF mRNA by interleukin 1 and tumor necrosis factor, *J. Exp. Med.* **167**: 1883-1893.

10. Gregory, H., J. Young, J.-M. Schröder, U. Mrowietz and E. Christophers. 1988. Structure determination of a human lymphocyte derived neutrophil activating peptide (LYNAP), *Biochem. Biophys. Res. Commun.* 151: 883-890.
11. Schröder, J.-M., U. Mrowietz and E. Christophers. 1988. Purification and partial biologic characterization of a human lymphocyte-derived peptide with potent neutrophil-stimulating activity, *J. Immunol.* 140: 3534-3540.
12. Van Damme, J., B. Decock, R. Conings, J.-P. Lenaerts, G. Opdenakker and A. Billiau. 1989. The chemotactic activity for granulocytes produced by virally infected fibroblasts is identical to monocyte-derived interleukin 8, Eur. *J. Immunol.* 19: 1189-1194.
13. Strieter, R.M., S.H. Phan, H.J. Showell, D.G. Remick, J.P. Lynch, M. Genord, C. Raiford, M. Eskandari, R.M. Marks and S.L. Kunkel. 1989. Monokine-induced neutrophil chemotactic factor gene expression in human fibroblasts. *J. Biol. Chem.* 264: 10621-10626.
14. Schröder, J.-M., M. Sticherling, H.H. Henneicke, W.G. Preissner and E. Christophers. 1990. IL-1α or tumor necrosis factor-α stimulate release of three NAP-1/IL-8-related neutrophil chemotactic proteins in human dermal fibroblasts, *J. Immunol.* 144: 2223-2232.
15. Schröder, J.M. and E. Christophers. 1989. Secretion of novel and homologous neutrophil-activating peptides by LPS-stimulated human endothelial cells, *J. Immunol.* 142: 244-251.
16. Strieter, R.M., S.L. Kunkel, H.J. Showell, D.G. Remick, S.H. Phan, P.A. Ward and R.M. Marks. 1989. Endothelial cell gene expression of a neutrophil chemotactic factor by TNF-α, LPS, and IL-1β, *Science* 243: 1467-1469.
17. Gimbrone Jr., M.A., M.S. Obin, A.F. Brock, E.A. Luis, P.E. Hass, C.A. Hbert, Y.K. Yip, D.W. Leung, D.G. Lowe, W.J. Kohr, W.C. Darbonne, K.B. Bechtol and J.B. Baker. 1989. Endothelial interleukin-8: a novel inhibitor of leukocyte-endothelial interactions, *Science* 246: 1601-1603.
18. Sica, A., K. Matsushima, J. Van Damme, J.M. Wang, N. Polentarutti, E. Dejana, F. Colotta and A. Mantovani. 1990. IL-1 transcriptionally activates the neutrophil chemotactic factor/IL-8 gene in endothelial cells, *Immunology* 69: 548-553.
19. Van Damme, J., R.A.D. Bunning, R. Conings, R. Graham, G. Russell and G. Opdenakker. 1990. Characterization of granulocyte chemotactic activity from human cytokine-stimulated chondrocytes as interleukin 8. *Cytokine* 2: 106-111.
20. Golds, E.E., P. Mason and P. Nyirkos. 1989. Inflammatory cytokines induce synthesis and secretion of gro protein and a neutrophil chemotactic factor but not β2-microglobulin in human synovial cells and fibroblasts, *Biochem. J.* 259: 585-588.
21. Larsen, C.G., A.O. Anderson, J.J. Oppenheim and K. Matsushima. 1989. Production of interleukin-8 by human dermal fibroblasts and keratinocytes in response to interleukin-1 or tumour necrosis factor, *Immunology* 68: 31-36.
22. Barker, J.N.W.N., V. Sarma, R.S. Mitra, V.M. Dixit and B.J. Nickoloff. 1990. Marked synergism between tumor necrosis factor-α and interferon- . in regulation of keratinocyte-derived adhesion molecules and chemotactic factors, *J. Clin. Invest.* 85: 605-608.
23. Elner, V.M., R.M. Strieter, S.G. Elner, M. Baggiolini, I. Lindley and S.L. Kunkel. 1990. Neutrophil chemotactic factor (IL-8) gene expression by cytokine-treated retinal pigment epithelial cells, *Am. J. Pathol.* 136: 745-750.
24. Kowalski, J. and D.T. Denhardt. 1989. Regulation of the mRNA for monocyte-derived neutrophil-activating peptide in differentiating HL60 promyelocytes, *Mol. Cell. Biol.* 9: 1946-1957.
25. Suzuki, K., H. Miyasaka, H. Ota, Y. Yamakawa, M. Tagawa, A. Kuramoto and S. Mizuno. 1989. Purification and partial primary sequence of a chemotactic protein for polymorphonuclear leukocytes derived from human lung giant cell carcinoma LU65C cells, *J. Exp. Med.* 169: 1895-1901.
26. Thornton, A.J., R.M. Strieter, I. Lindley, M. Baggiolini and S.L. Kunkel. 1990. Cytokine-induced gene expression of a neutrophil chemotactic factor/IL-8 in human hepatocytes, *J. Immunol.* 144: 2609-2613.
27. Van Damme, J., B. Decock, J.-P. Lenaerts, R. Conings, R. Bertini, A. Mantovani and A. Billiau. 1989. Identification by sequence analysis of chemotactic factors for monocytes produced by normal and transformed cells stimulated with virus, double-stranded RNA or cytokine, *Eur. J. Immunol.* 19: 2367-2373.
28. Yoshimura, T., N. Yuhki, S.K. Moore, E. Appella, M.I. Lerman and E.J. Leonard. 1989. Human monocyte chemoattractant protein-1 (MCP-1). *FEBS Letts.* 244: 487-493.
29. Furutani, Y., H. Nomura, M. Notake, Y. Oyamada, T. Fukui, M. Yamada, C.G. Larsen, J.J. Oppenheim and K. Matsushima. 1989. Cloning and sequencing of the cDNA for human monocyte chemotactic and activating factor (MCAF), *Biochem. Biophys. Res. Commun.* 159: 249-255.
30. Decock, B., R. Conings, J.-P. Lenaerts, A. Billiau and J. Van Damme. 1990. Identification of the monocyte chemotactic protein from human osteosarcoma cells and monocytes: detection of a novel N-terminally processed form, *Biochem. Biophys. Res. Commun.* 167: 904-909.

31. Yoshimura, T., E.A. Robinson, S. Tanaka, E. Appella and E.J. Leonard. 1989. Purification and amino acid analysis of two human monocyte chemoattractants produced by phytohemag-glutinin-stimulated human blood mononuclear leukocytes, *J. Immunol.* **142:** 1956-1962.
32. Rollins, B.J., P. Stier, T. Ernst and G.G. Wong. 1989. The human homolog of the *JE* gene encodes a monocyte secretory protein, *Mol. Cell. Biol.* 9:4687-4695.
33. Introna, M., R.C. Bast Jr., C.S. Tannenbaum, T.A. Hamilton and D.O. Adams. 1987. The effect of LPS on expression of the early "competence" genes *JE* and *KC* in murine peritoneal macrophages, *J. Immunol.* **138:** 3891-3896.
34. Prpic, V., S.-F. Yu, F. Figueiredo, P.W. Hollenbach, G. Gawdi, B. Herman, R.J. Uhing and D.O. Adams. 1989. Role of Na+/H+ exchange by interferon- . in enhanced expression of *JE* and I-Aβ genes, *Science* **244:** 469-471.
35. Rollins, B.J., E.D. Morrison, and C.D. Stiles. 1988. Cloning and expression of JE, a gene inducible by platelet-derived growth factor and whose product has cytokine-like properties, *Proc. Natl. Acad. Sci. USA* **85:** 3738-3742.
36. C.G. Larsen, C.O.C. Zachariae, J.J. Oppenheim and K. Matsushima. 1989. Production of monocyte chemotactic and activating factor (MCAF) by human dermal fibroblasts in response to interleukin 1 or tumor necrosis factor, *Biochem. Biophys. Res. Commun.* **160:** 1403-1408.
37. Strieter, R.M., R. Wiggins, S.H. Phan, B.L. Wharram, H.J. Showell, D.G. Remick, S.W. Chensue and S.L. Kunkel. 1989. Monocyte chemotactic protein gene expression by cytokine-treated human fibroblasts and endothelial cells, *Biochem. Biophys. Res. Commun.* **162:** 694-700.
38. Sica, A., J.M. Wang, F. Colotta, E. Dejana, A. Mantovani, J.J. Oppenheim, C.G. Larsen, C.O.C. Zachariae and K. Matsushima. 1990. Monocyte chemotactic and activating factor gene expression induced in endothelial cells by IL-1 and tumor necrosis factor, *J. Immunol.* **144:** 3034-3038.
39. Takehara, K., E.C. LeRoy and G.R. Grotendorst. 1987. TGF-β inhibition of endothelial cell proliferation: alteration of EGF binding and EGF-induced growth-regulatory (competence) gene expression, *Cell* **49:** 415-422.
40. Valente, A.J., D.T. Graves, C.E. Vialle-Valentin, R. Delgado and C.J. Schwartz. 1988. Purification of a monocyte chemotactic factor secreted by nonhuman primate vascular cells in culture, *Biochemistry* **27:** 4162-4168.
41. Matsushima, K., C.G. Larsen, G.C. DuBois and J.J. Oppenheim. 1989. Purification and characterization of a novel monocyte chemotactic and activating factor produced by a human myelomonocytic cell line, *J. Exp. Med.* **169:** 1485-1490.
42. Graves, D.T., Y.L. Jiang, M.J. Williamson and A.J. Valente. 1989. Identification of monocyte chemotactic activity produced by malignant cells, *Science* **245:** 1490-1493.
43. Bottazzi, B., F. Colotta, A. Sica, N. Nobili and A. Mantovani. 1990. A chemoattractant expressed in human sarcoma cells (tumor-derived chemotactic factor, TDCF) is identical to monocyte chemoattractant protein-1/monocyte chemotactic and activating factor (MCP-1/MCAF), *Int. J. Cancer* **45:** 795-797.
44. Yoshimura, T., E.A. Robinson, S. Tanaka, E. Appella, J.-I. Kuratsu and E.J. Leonard. 1989. Purification and amino acid analysis of two human glioma-derived monocyte chemoattractants, *J. Exp. Med.* **169:** 1449-1459.
45. Nelson, R.D., P.G. Quie and R.L. Simmons. 1975. Chemotaxis under agarose: a new and simple method for measuring chemotaxis and spontaneous migration of human polymorphonuclear leukocytes and monocytes, *J. Immunol.* **115:** 1650-1656.
46. Westwick, J., S.W. Li and R.D. Camp. 1989. Novel neutrophil-stimulating peptides, *Immunol. Today* **10:** 146-147.
47. Van Damme, J., M.R. Schaafsma, W.E. Fibbe, J.H.F. Falkenburg, G. Opdenakker and A. Billiau. 1989. Simultaneous production of interleukin 6, interferon-β and colony-stimulating activity by fibroblasts after viral and bacterial infection. *Eur. J. Immunol.* **19:** 163-168.
48. Strieter, R.M., S.W. Chensue, T.J. Standiford, M.A. Basha, H.J. Showell and S.L. Kunkel. 1990. Disparate gene expression of chemotactic cytokines by human mononuclear phagocytes, *Biochem. Biophys. Res. Commun.* **166:** 886-891.
49. Rampart, M., J. Van Damme, L. Zonnekeyn and A.G. Herman. 1989. Granulocyte chemotactic protein/interleukin-8 induces plasma leakage and neutrophil accumulation in rabbit skin. *Am. J. Pathol.* **135:** 21-25.
50. Bottazzi, B., N. Polentarutti, R. Acero, A. Balsari, D. Boraschi, P. Ghezzi, M. Salmona and A. Mantovani. 1983. Regulation of the macrophage content of neoplasms by chemoattractants, *Science* **220:** 210 – 212.

ACTIVATION OF HUMAN NEUTROPHILS BY NAP-1 AND OTHER CHEMOTACTIC AGONISTS

Marco Baggiolini, Beatrice Dewald, and Alfred Walz

Theodor-Kocher Institute
University of Bern
P.O. Box 99, CH-3000 Bern 9, Switzerland

INTRODUCTION

The neutrophils normally circulate passively in the blood and are recruited at sites of infection or inflammation by chemotactic agonists that turn them on and direct their migration into the tissues. The adhesion to the venular endothelium, the passage through the vessel wall and the directed movement in the interstitium are complex events involving motile and secretory functions. Three main responses of neutrophils, the shape change, the exocytosis of storage components, and the respiratory burst can be analyzed in real time *in vitro*, and we shall describe here how these responses are induced by NAP-1 and other chemotactic agonists.

THE SHAPE CHANGE

The shape change reflects the activation of the motile apparatus. It is measured by nephelometry in diluted neutrophil suspensions and can be quantified according to Beer's law as a transient increase in suspension transmission when a collimated laser beam is used as the light source and forward-scattered light is prevented from reaching the detector with a beam stop (1). The transmission increase is due to the protrusion of cytoplasmic lamellae and the consequent decrease in cell body volume. In stirred suspensions of 0.8 x 10^6 cells per ml in a square cuvette with 1 cm light path the overall transmission change is in the range of 4-8%. The transmission rises rapidly upon agonist stimulation, reaches a maximum in 20-30 sec and returns to the initial level within a few min. The neutrophil shape change responses to NAP-1 and C5a are practically identical in extent and last for 1.5 to 2 min. The effect of fMet-Leu-Phe is similar in extent, but of longer duration (3-4 min) (2). When the neutrophils are pretreated with phorbol myristate acetate (PMA) or with the fungal metabolite 17-hydroxy-wortmannin and then stimulated with a chemotactic receptor agonist, the transmission changes show regular oscillations with a period of about 8 sec. The characteristics of the oscillatory shape response is the same following stimulation with C5a, fMet-Leu-Phe, PAF and LTB_4 (1), and, as shown by later experiments, NAP-1. According to the light scattering theory, such oscillations are indicative of rhythmic changes in cell body volume presumably related to the protrusion and retraction of lamellipodia (1).

Chemotactic Cytokines, Edited by J. Westwick *et al.*
Plenum Press, New York, 1991

EXOCYTOSIS

The Granules

Human neutrophils contain two types of granules which can be distinguished unequivocally in myeloperoxidase-stained transmission electron micrograph where azurophil granules are positive and specific granules negative. Both granules can be separated by centrifugation owing to their difference in density and mass (3,4). They differ almost completely in their composition, their only common constituent being lysozyme. The azurophil granules are rich in lytic enzymes including the acid hydrolases that are characteristic for lysosomes, lysozyme, and three serine proteases with a neutral pH optimum. Myeloperoxidase is the only non-hydrolytic enzyme. In terms of actual protein, elastase, cathepsin G, lysozyme and myeloperoxidase are the major constituents. The specific granules contain collagenase, lysozyme, lactoferrin and vitamin B_{12}-binding proteins. While it is clear from their enzymic equipment that the azurophil granules are involved in both microbial killing and digestion of biological material, it is more difficult to speculate about the possible function of the specific granules. It is hoped that the discovery of new components will soon help to understand the functional role of these granules which constitute the largest storage compartment of the mature neutrophil.

Subcellular fractionation studies and exocytosis experiments have demonstrated the existence of smaller storage organelles that are unrelated to the granules. These structures contain some acid hydrolases, including a relatively large proportion of cathepsin B and D and of proteinase 3, which also occur in the azurophil granules, and gelatinase (5,6). Such organelles cannot be positively identified in the microscope and, therefore, it has not been possible to establish the mode and time of their formation. In a recent study, Hibbs and Bainton (7) described the immunohistochemical co-localization of gelatinase and lactoferrin, suggesting that gelatinase is a component of the specific granules. These conclusions contrast with the results of our own studies which revealed a dissociation between gelatinase and specific granule components (5,6). Other subcellular compartments that are subject to exocytosis were reported to contain alkaline phosphatase (8) and, as recently shown, tetranectin (9). Small organelles, and in some instances the specific granules, are considered as the stores of receptors, e.g. CR1, CR3 and the fMet-Leu-Phe receptor (10-12), and of adhesion proteins (13,14), whose expression on the phagocyte surface is enhanced upon stimulation. The membrane of the specific granules appears to function as a an intracellular store of cytochrome b_{558}, the terminal element of the respiratory burst oxidase (15).

Release of Storage Proteins

The release of granule enzymes independent of phagocytosis was first reported to Woodin and Wieneke who used leukocidin as a stimulus. About ten years later, Estensen and colleagues showed that phorbol myristate acetate (PMA) induces the concentration-dependent release of lysozyme, but little or no release of β-glucuronidase and myeloperoxidase, and concluded that selective discharge of specific granules was taking place. The effect of PMA was confirmed by Goldstein *et al.* and Wright *et al.* who showed that stimulation with PMA results in a depletion of specific granules within the cells (see (16) for original references). With the identification and purification of chemotactic peptides and chemotactic lipids (see (16) for references) it became clear that all these agonists induce exocytosis of specific granules and smaller storage organelles, but not of azurophil granules (16). In addition to the release of enzymes and other soluble components, stimulation with chemotactic agonists results in the expression of proteins that are associated with the membrane of intracellular storage organelles, like receptors and cytochrome

b_{558} on the plasmalemmal surface (see above). NAP-1 behaves exactly as the known chemotaxins like C5a and fMet-Leu-Phe: In normal human neutrophils it induces the release of vitamin B12-binding protein and gelatinase (but not of elastase, β-glucuronidase or myeloperoxidase) (17), and induces the upregulation of CR1 (18) and CR3 (19).

Pretreatment with cytochalasin B renders the neutrophils more responsive to chemotactic agonists, the release of specific granules components and that of gelatinase are enhanced, and marked exocytosis of azurophil granule contents is observed in addition (16). The mechanism of action of cytochalasin B is unknown, but one can take advantage of its enhancing effect for the study of agents or conditions that influence neutrophil exocytosis. In fact, NAP-1 was found in our laboratory on the basis of its ability to induce elastase release from cytochalasin B-treated human neutrophils (17,20).

THE RESPIRATORY BURST

The respiratory burst is the most characteristic response of phagocytes to stimulation. It is due to the activation of the NADPH:O_2^- oxidoreductase (NADPH-oxidase), a transmembrane electron transport chain with cytosolic NADPH as the electron donor and oxygen as the acceptor. The product of the oxidase is O_2^-. As shown by the reaction sequence:

$$4NADPH + 4O_2 \rightarrow 4H^+ + 4O_2^- \rightarrow 2O_2 + 2H_2O_2 \rightarrow 3O_2 + 2H_2O$$

O_2^- and H_2O_2 are subject to dismutation. When side reactions are excluded, for each mol of oxygen that is reduced, 1 mol of O_2^- or ½ mol of H_2O_2 is recovered. Since 3/4 of the oxygen is regenerated, only ¼ of the actual consumption is detected. The activity of the oxidase can be assessed as the superoxide dismutase-sensitive reduction of cytochrome c which is a measure of O_2^- (21,22), or the oxidation of chromogenic, fluorogenic or luminogenic H-donors catalyzed by added peroxidase, which is a measure of H_2O_2 (23-26). Under rigorous conditions, luminol-dependent chemiluminescence is proportional to the rate of H_2O_2 formation and thus directly reflects the activity of the NADPH-oxidase (26). Non-mitochondrial oxygen consumption may also be used as an indirect measurement of the respiratory burst.

Oxidase Components

In resting phagocytes the NADPH-oxidase is disassembled and its components are located in the plasma membrane, the ectoplasmic cytoskeleton and the cytosol (27-29). The main membrane-associated component is a b-type cytochrome, termed cytochrome b_{558}, because of its absorption maximum at $_{558}$nm (15,30). The cytochrome is a heterodimer consisting of 92kDa and 22kDa subunits, which were recently cloned (31,32). The heme is bound to the small subunit (33), has an unusually low midpoint potential ($E_{m.7.0}$ -245mV) (34) and reduces molecular oxygen to superoxide ($E_{m.7.0}$ -160 mV) (35). A low molecular weight (22kDa), *ras*-related GTP-binding protein, termed rap-1, is associated with the cytochrome (36). The membrane also contains a 66kDa flavoprotein as intermediate electron carrier (37,38), a proton channel, which possibly compensates for the vectorial electron transport through the oxidase (39,40), and a 45kDa flavoprotein which binds the NADPH-oxidase inhibitor diphenyl iodonium (41,42) are also membrane bound. Four cytosolic components (commonly called cytosolic factors) with isoelectric points of 3.1, 6.1, 7 and 10 (43), have been identified. Two of them, p47 and p67, have been cloned and found to substitute for the cytosol in the cell-free activation of the NADPH-oxidase (44). p47 is a protein kinase C substrate (45), and is strongly phosphorylated and translocated to the plasma membrane together with p67 in intact neutrophils upon stimulation (46–48). A 66kDa NADPH-binding protein was also identified in the cytosol (49–51). On stimulation, this additional cytosolic factor could associate with the membrane and become part of the electron transport chain from NADPH to oxygen.

The study of chronic granulomatous disease (CGD), a group of inherited disorders collectively characterized by an impairment of the respiratory burst (52), has shown that several of the components described are essential for the activity of the NADPH-oxidase. There are in fact forms of CGD due to defects of the β (31,53) or the α (54) subunit of the cytochrome b_{558}, and others due to the lack of either the p47 or the p67 cytosolic factor (53).

Activation of the Respiratory Burst

In view of the biochemical complexity and the need to assemble components from different subcellular compartments, it is astonishing to realize that upon stimulation with chemotactic agonist the oxidase is turned on in about 2 sec. Wymann *et al.* (55) have recently performed a kinetic study of the activation of the respiratory burst by fMet-Leu-Phe, C5a, PAF and LTB_4 using a chemiluminescence assay for H_2O_2, which combines high sensitivity with high signal-to-noise ratio (26). Despite major differences in intensity and duration, the responses to maximum-effective concentrations of the four agonists and different concentrations of fMet-Leu-Phe show similar kinetic features. With all four agonists, the onset time of H_2O_2 production is identical, and the slope of the chemiluminescence progress curves at the first inflection point extrapolates back to the same point on the time axis. This suggests that stimuli acting through distinct receptors are transduced by a common mechanism, and that the velocity of the signal transduction process leading to activation of the NADPH-oxidase operates is intrinsic rather than agonist-dependent. In contrast to the onset and the initial phase, the duration of the respiratory burst varies considerably with the agonist.

For reasons that still have to be investigated, the responses to PAF and LTB_4 are turned off much more rapidly than those induced by fMet-Leu-Phe or C5a. NAP-1 is no exception to this rule: It induces a response with rapid onset (about 2 sec) and high rate of activation, which is intermediate in terms of extent and duration between those of fMet-Leu-Phe and C5a and those of the lipid mediators.

A respiratory burst can also be induced with phorbol esters, permeant diacylglycerols or calcium ionophores (55,56). These stimuli, however, act more slowly than chemotactic agonists: The onset times of the responses are at least 4 to 5-fold longer and the apparent rates of generation of active oxidase are considerably lower (55).

SIGNAL TRANSDUCTION EVENTS

All three main responses of neutrophils are observed upon stimulation by anyone of the chemotactic agonists described, including NAP-1. Although structurally unrelated and acting via distinct receptors, these stimuli initiate common signal transduction events and appear to elicit the cellular responses by similar mechanisms. The responses themselves are controlled at different levels of the signal transduction process, and the control is particularly tight for the respiratory burst, which is understandable in view of the toxic effects of its products. We shall give a description of the signal transduction events required for the induction of the respiratory burst and discuss later the differences with respect to shape change and exocytosis.

Receptor Occupancy

The respiratory burst is initiated by the binding of the agonist to its receptor and decays rapidly when the agonist is displaced by an antagonist or by binding with an antibody (57). These observations indicate that persistence of the agonist-receptor complex is necessary to maintain the oxidase in its active form and suggest that the active oxidase is labile and may be assembled and disassembled in a cyclic process as long as the receptor remains occupied. For a given agonist, the intensity of the response is related to the number of receptors occupied. With N-formyl peptides maximum rates of superoxide production are reached at approximately 10% occupancy (57). A much lower receptor occupancy is required for other responses, namely chemotaxis and the release of some enzymes.

GTP-binding Proteins

The involvement of GTP-binding proteins in receptor-mediated neutrophil activation was originally suggested by the blocking effect of pretreatment of the cells with *Bordetella pertussis* toxin (58). The toxin inactivates GTP-binding proteins of the inhibitory (G_i) type through the ADP-ribosylation of their α-subunit preventing association with the agonist receptor (59). Further evidence came from the observation that neutrophil responses are induced and/or enhanced by fluoride (60) and, in electro-permeabilized cells, by the non-hydrolyzable GTP analogue, GTP-gammaS (61). Human neutrophils possess a GTP-binding protein with a distinct α subunit, G_{i2}-α, which is linked to myristic acid at its amino terminus (62). This lipophilic adduct may ensure the association of the α-subunit with the plasma membrane (63,64). Upon pertussis toxin treatment the neutrophils become unresponsive to chemotactic agonists, but the respiratory burst is still perfectly inducible with stimuli that bypass receptors like phorbol esters, permeant diacylglycerols or calcium ionophores.

The Formation of Second Messengers

Upon receptor-ligand interaction and coupling of the G-protein to the receptor a phospholipase C is activated that generates two second messengers, 1,4,5 inositol trisphosphate (IP_3) and diacylglycerol, through the cleavage of phosphatidyl-inositol 4,5-bisphosphate (65). IP_3 is released into the cytosol and binds to specific receptors on intracellular calcium storage organelles (66) inducing a rise in cytosolic free calcium (67). Diacylglycerol, remains associated with the membrane and activates protein kinase C. Recently, stimulus-dependent activation of phospholipase D generating phosphatidic acid, which is then dephosphorylated, was described as an additional source of diacylglycerol (68).

The Role of Cytosolic Free Calcium

Although it is well known that chemotactic agonists, including NAP-1, induce a rapid and transient increase of cytosolic free calcium (2,17,67), the actual molecular role of this cation in neutrophil responses has not been clarified. Real-time recordings show that the agonist-induced calcium rise is preceded by a short lag which is thought to depend the time required for the generation of active concentrations of IP_3 by phospholipase C (69). The calcium change reflects the liberation from the intracellular storage pool and the influx through the plasma membrane. Calcium influx, however, is not essential for signal transduction since all neutrophil responses can be elicited in the absence of extracellular calcium (67) provided that the intracellular stores are not depleted. IP_3 does not appear to act on the plasma membrane, and it has been suggested on the basis of patch-clamp experiments that calcium may enter stimulated neutrophils through calcium-dependent cation channels (69).

Neutrophils can be depleted of mobilizable intracellular calcium through the combined use of a calcium ionophore and intra- and extracellular chelators (70). In such cells, the resting calcium level is below normal, and no rise can be induced by stimulation indicating that the mobilisable storage pool is indeed depleted (71). In addition phospholipase C activation is prevented as indicated by the lack of formation of IP_3 and diacylglycerol (70). Under these conditions receptor agonist stimulation does not lead to activation of protein kinase C and of the NADPH-oxidase. A respiratory burst response is obtained, however, when protein kinase C is activated by phorbol esters or exogenous diacylglycerol (70,71). These observations indicate that the kinase is required for the induction of the respiratory burst. Calcium depletion prevents the burst presumably because under these conditions phospholipase C remains inactive and no diacylglycerol is formed. In contrast to the formation of diacylglycerol, the transient rise in intracellular calcium does not appear to be essential since the respiratory burst can be induced with receptor agonists even in calcium-depleted cells, provided that phorbol esters or similar agents are added to activate protein kinase C.

Protein Kinase C

The respiratory burst can be induced with phorbol myristate acetate (72) and permeant diacylglycerols (56) suggesting that protein kinase C is part of the signal transduction sequence. A role of protein kinase C and protein phosphorylations is also suggested by the effect protein kinase inhibitors like staurosporine and sphingosine bases, which prevent phosphorylation and attenuate the respiratory burst response to protein kinase C activators and receptor agonists (73,74). The kinetics of the respiratory burst induced by phorbol esters and diacylglycerols on the one hand and chemotactic agonists on the other, however, differ substantially. As shown by high resolution recordings of the H_2O_2 production (55) the onset of the response is much more rapid, and the apparent rate of activation of the NADPH-oxidase considerably higher when an agonist is used as the stimulus, indicating that the signal transduction initiated by chemotactic receptor ligation involves other processes in addition to protein kinase activation.

Other Signal Transduction Events

In human neutrophils stimulated with a chemotactic agonist a distinct lag is observed prior to calcium mobilization (69). This lag is shorter than the onset time of the respiratory burst (55) and, therefore, the cytosolic free calcium rises before the NADPH-oxidase is activated. The onset time of the burst, however, is markedly shortened following pretreatment with activators of protein kinase C. When neutrophils are activated with a low concentration of phorbol myristate acetate and then stimulated with a chemotactic agonist they respond with a major rise in H_2O_2 production that ensues without a measurable lag and clearly precedes the rise in cytosolic free calcium. This experiment reveals a receptor-mediated respiratory burst response that is calcium-independent (55). Such a signal transduction event has also been evidenced in calcium-depleted cells which rapidly respond to agonists (despite their inability to mobilize calcium) provided that protein kinase C is turned on with a phorbol ester (70). Further evidence comes from the use of 17-hydroxy-wortmannin, a fungal metabolite that blocks the respiratory burst induced by chemotactic agonists or phagocytosis (71,75). In calcium-depleted neutrophils pretreated with threshold concentrations of phorbol myristate acetate, 17-hydroxy-wortmannin prevents the (calcium-independent) response to chemotactic agonists. 17-Hydroxy-wortmannin has, on the other hand, no influence on calcium mobilization in normal cells and does not modify the function of G-proteins, phospholipase C and protein kinase C (71).

These observations indicate that the activation of the respiratory burst by chemotactic agonists depends on two distinct signal transduction processes. One is calcium-dependent and leads to the activation of protein kinase C, while the other is calcium- and protein kinase C-independent. Since the respiratory burst response to agonists can be prevented by either calcium depletion or treatment with 17-hydroxy-wortmannin, it appears that both processes must be acting in concert.

RESPONSES INDUCED BY NAP-1 AND DIFFERENT LEVELS OF SIGNAL TRANSDUCTION CONTROL

As outlined, the profile of activity of NAP-1 on human neutrophils is qualitatively identical to that observed with C5a and fMet-Leu-Phe, and this means that NAP-1 fully qualifies as a neutrophil chemotactic agonist. NAP-1 is also active in vivo: When applied locally, it induces massive and long-lasting neutrophil recruitment and plasma exudation (76). Shape change, exocytosis and the respiratory burst are controlled at different levels of the signal transduction process, and also in this respect there is complete correspondence between NAP-1 and the other chemotactic agonists.

Effects of Inhibitors and Calcium Depletion

The validity of the signal transduction process involved in the initiation of the respiratory burst can be tested with a number of manipulations known to interfere with single events. Similar experiments serve to identify differences in the regulation of the other responses, shape change and exocytosis, as briefly summarized here. Two elements of the signal transduction chain, the receptor and a GTP-binding protein are necessary for all responses. The possibility to prevent activation with antagonists has been documented for fMet-Leu-Phe (57) and PAF (77). The ligated receptor must be connected to the transduction machinery through a GTP-binding protein. In neutrophils these proteins are inactivated by *pertussis* toxin, and in fact such a pretreatment inhibits all responses to NAP-1 and other chemotaxins (2).

Depletion of mobilizable cytosolic calcium reveals that shape change and product release are controlled in different ways. Calcium-depleted neutrophils undergo a perfectly normal shape change and shape change oscillations (78,79) induced by chemotactic agonists including NAP-1, but lack exocytosis and the respiratory burst. The shape change elicited by NAP-1 is also unaffected by 17-hydroxy-wortmannin and by staurosporine (2,78) indicating that the calcium-independent signal transduction pathway and protein kinase C are not required for activating the motor apparatus of the cells.

The use of the two latter inhibitors shows in addition that granule exocytosis and the respiratory burst are under different control. Both responses are inhibited by 17-hydroxy-wortmannin (2,71) while staurosporine only inhibits the burst (2,74). Staurosporine does not affect the release of azurophil granule enzymes and actually acts as a stimulus of specific granule exocytosis (74), suggesting that phosphorylation reactions may have a negative effect on the release process.

REFERENCES

1. Wymann, M.P., P. Kernen, D.A. Deranleau, B. Dewald, V. von Tscharner, and M. Baggiolini. 1987. Oscillatory motion in human neutrophils responding to chemotactic stimuli. *Biochem. Biophys. Res. Commun.* **147**: 361-368.

2. Thelen, M., P. Peveri, P. Kernen, V. von Tscharner, A. Walz, and M. Baggiolini. 1988. Mechanism of neutrophil activation by NAF, a novel monocyte-derived peptide agonist. *FASEB. J.* **2**: 2702-2706.
3. Bretz, U., and M. Baggiolini. 1974. Biochemical and morphological characterization of azurophil and specific granules of human neutrophilic polymorphonuclear leukocytes. J.Cell Biol. **63**: 251-269.
4. Spitznagel, J.K., F. Dallegri, M.S. Leffell, J.D. Folds, I.R.H. Welsh, M.H. Cooney, and L.E. Martin. 1974. Character of azurophil and specifis granules purified from human polymorphonuclear leukocytes. *Lab. Invest.* **30**: 774-785.
5. Baggiolini, M., J. Schnyder, U. Bretz, B. Dewald, and W. Ruch. 1980. Cellular mechanisms of proteinase release from inflammatory cells and the degradation of extracellular proteins. *Ciba. Found. Symp.* **75**: 105-121.
6. Dewald, B., U. Bretz, and M. Baggiolini. 1982. Release of gelatinase from a novel secretory compartment of human neutrophils. *J. Clin. Invest.* **70**: 518-525.
7. Hibbs, M.S., and D.F. Bainton. 1989. Human neutrophil gelatinase is a component of specific granules. *J. Clin. Invest.* **84**: 1395-1402.
8. Borregaard, N., L.J. Miller, and T.A. Springer. 1987. Chemoattractant-regulated mobilization of a novel intracellular compartment in human neutrophils. *Science* **237**: 1204-1206.
9. Borregaard, N., L. Christensen, O.W. Bjerrum, H.S. Birgens, and I. Clemmensen. 1990. Identification of a highly mobilizable subset of human neutrophil intracellular vesicles that contains tetranectin and latent alkaline phosphatase. *J. Clin. Invest.* **85**: 408-416.
10. Fearon, D.T., and L.A. Collins. 1983. Increased expression of C3b receptors on polymorphonuclear leukocytes induced by chemotactic factors and by purification procedures. J.Immunol.**130**: 370-375.
11. Fletcher, M.P., B.E. Seligmann, and J.I. Gallin. 1982. Correlation of human neutrophil secretion, chemoattractant receptor mobilization and enhanced functional capacity. *J. Immunol.* **128**: 941-948.
12. O Shea, J.J., E.J. Brown, B.E. Seligmann, J.A. Metcalf, M.M. Frank, and J.I. Gallin. 1985. Evidence for distinct intracellular pools of receptors C3b and C3bi in human neutrophils. *J. Immunol.* **134**: 2580-2587.
13. Bainton, D.F., L.J. Miller, T.K. Kishimoto, and T.A. Springer. 1987. Leukocyte adhesion receptors are stored in peroxidase-negative granules of human neutrophils. *J. Exp. Med.* **166**: 1641-1653.
14. Singer, I.I., S. Scott, D.W. Kawka, and D.M. Kazazis. 1989. Adhesomes: specific granules containing receptors for laminin, C3bi/fibrinogen, fibronectin and vitronectin in human polymorphonuclear leukocytes and monocytes. *J. Cell Biol.* **109**: 3169-3182.
15. Segal, A.W., and O.T.G. Jones. 1979. The subcellular distribution and some properties of the cytochrome b component of the microbicidal oxidase system of human neutrophils. *Biochem. J.* **182**: 181-188.
16. Baggiolini, M., and B. Dewald. 1984. Exocytosis by neutrophils. *Contemp. Top. Immunobiol.* **14**: 221-246.
17. Peveri, P., A. Walz, B. Dewald, and M. Baggiolini. 1988. A novel neutrophil-activating factor produced by human mononuclear phagocytes. *J. Exp. Med.* **167**: 1547-1559.
18. Paccaud, J.-P., J.A. Schifferli, and M. Baggiolini. 1990. NAP-1/IL-8 induces upregulation of CR1 receptors in human neutrophil leukocytes. *Biochem. Biophys. Res. Commun.* **166**: 187-192.
19. Detmers, P.A., S.K. Lo, E. Olsen-Egbert, A. Walz, M. Baggiolini, and Z.A. Cohn. 1990. Neutrophil-activating protein 1/interleukin 8 stimulates the binding activity of the leukocyte adhesion receptor CD11b/CD18 on human neutrophils. *J. Exp. Med.* **171**: 1155-1162.
20. Walz, A., P. Peveri, H. Aschauer, and M. Baggiolini. 1987. Purification and amino acid sequencing of NAF, a novel neutrophil-activating factor produced by monocytes. *Biochem. Biophys. Res. Commun.* **149**: 755-761.
21. Markert, M., P.C. Andrews, and B.M. Babior. 1984. Measurement of O_2-production by human neutrophils. The preparation and assay of NADPH oxidase-containing particles from human neutrophils. *Methods. Enzymol.* **105**: 358-365.
22. Thelen, M., M. Wolf, and M. Baggiolini. 1988. Activation of monocytes by interferon-gamma has no effect on the level or affinity of the nicotinamide adenine dinucleotide-phosphate oxidase and on agonist-dependent superoxide formation. *J. Clin. Invest.* **81**: 1889-1895.
23. Pick, E., and D. Mizel. 1981. Rapid microassays for the measurement of superoxide and hydrogen peroxide production by macrophages in culture using an automatic enzyme immunoassay reader. *J. Immunol. Methods.* **46**: 211-226.

24. Hyslop, P.A., and L.A. Sklar. 1984. A quantitative fluorimetric assay for the determination of oxidant production by polymorphonuclear leukocytes: its use in the simultaneous fluorimetric assay of cellular activation processes. *Anal. Biochem.* 141: 280-286.
25. Ruch, W., P.H. Cooper, and M. Baggiolini. 1983. Assay of H_2O_2 production by macrophages and neutrophils with homovanillic acid and horse-radish peroxidase. *J. Immunol.* Methods. 63: 347-357.
26. Wymann, M.P., V. von Tscharner, D.A. Deranleau, and M. Baggiolini. 1987. Chemiluminescence detection of H_2O_2 produced by human neutrophils during the respiratory burst. *Anal. Biochem.* 165: 371-378.
27. McPhail, L.C., P.S. Shirley, C.C. Clayton, and R. Snyderman. 1985. Activation of the respiratory burst enzyme from human neutrophils in a cell-free system. Evidence for a soluble cofactor. *J. Clin. Invest.* 75: 1735-1739.
28. Curnutte, J.T. 1985. Activation of human neutrophil nicotinamide adenine dinucleotide phosphate, reduced (triphosphopyridine nucleotide, reduced) oxidase by arachidonic acid in a cell-free system. *J. Clin. Invest.* 75: 1740-1743.
29. Babior, B.M., J.T. Curnutte, and N. Okamura. 1988. The respiratory burst oxidase and certain members of the 48kD phosphoprotein familly are associated with the neutrophil cytoskeleton. *Blood* 72 (5), suppl.1: 141a-141a.
30. Segal, A.W., and O.T. Jones. 1978. Novel cytochrome b system in phagocytic vacuoles of human granulocytes. *Nature* 276: 515-517.
31. Royer Pokora, B., L.M. Kunkel, A.P. Monaco, S.C. Goff, P.E. Newburger, R.L. Baehner, F.S. Cole, J.T. Curnutte, and S.H. Orkin. 1986. Cloning the gene for an inherited human disorder - chronic granulomatous disease - on the basis of its chromosomal location. *Nature* 322: 32-38.
32. Parkos, C.A., M.C. Dinauer, L.E. Walker, R.A. Allen, A.J. Jesaitis, and S.H. Orkin. 1988. Primary structure and unique expression of the 22-kilodalton light chain of human neutrophil cytochrome b. *Proc. Natl. Acad. Sci. USA* 85: 3319-3323.
33. Nugent, J.H.A., W. Gratzer, and A.W. Segal. 1989. Identification of the haem-binding subunit of cytochrome b_{245}. *Biochem. J.* 264: 921-924.
34. Cross, A.R., O.T. Jones, A.M. Harper, and A.W. Segal. 1981. Oxidation-reduction properties of the cytochrome-b found in the plasma-membrane fraction of human neutrophils: a possible oxidase in the respiratory burst. *Biochem. J.* 194: 599-607.
35. Prince, R.C., and D.E. Gunson. 1987. Superoxide production by neutrophils. *Trends Biochem. Sci.* 12: 86-87.
36. Quinn, M.T., C.A. Parkos, L. Walker, S.H. Orkin, M.C. Dinauer, and A.J. Jesaitis. 1989. Association of a Ras-related protein with cytochrome b of human neutrophils. *Nature* 342: 198-200.
37. Doussiere, J., and P.V. Vignais. 1985. Purification and properties of an O_2-.-generating oxidase from bovine polymorphonuclear neutrophils. *Biochemistry.* 24: 7231-7239.
38. Markert, M., G.A. Glass, and B.M. Babior. 1985. Respiratory burst oxidase from human neutrophils: purification and some properties. *Proc. Natl. Acad. Sci. USA* 82: 3144-3148.
39. Henderson, L.M., J.B. Chappell, and O.T. Jones. 1987. The superoxide-generating NADPH oxidase of human neutrophils is electrogenic and associated with an H^+ channel. *Biochem. J.* 246: 325-329.
40. Henderson, L.M., J.B. Chappell, and O.T. Jones. 1988. Superoxide generation by the electrogenic NADPH oxidase of human neutrophils is limited by the movement of a compensating charge. *Biochem. J.* 255: 285-290.
41. Cross, A.R., and O.T. Jones. 1986. The effect of the inhibitor diphenylene iodonium on the superoxide-generating system of neutrophils. Specific labelling of a component polypeptide of the oxidase. *Biochem. J.* 237: 111-116.
42. Yea, C.M., A.R. Cross, and O.T.G. Jones. 1990. Purification and some properties of the 45 kDa diphenylene iodonium-binding flavoprotein of neutrophil NADPH oxidase. *Biochem. J.* 265: 95-100.
43. Curnutte, J.T., P.J. Scott, and L.A. Mayo. 1989. Cytosolic components of the respiratory burst oxidase: Resolution of four components, two of which are missing in complementing types of chronic granulomatous disease. *Proc. Natl. Acad. Sci. USA* 86: 825-829.
44. Leto, T.L., K.J. Lomax, B.D. Volpp, H. Nunoi, J.M.G. Sechler, W.M. Nauseef, R.A. Clark, J.I. Gallin, and H.L. Malech. 1990. Cloning of a 67-kD neutrophil oxidase factor with similarity to a noncatalytic region of p60c-src. *Science* 248: 727-730.

45. Kramer, I.M., A.J. Verhoeven, Bend.R.L. van der, R.S. Weening, and D. Roos. 1988. Purified protein kinase C phosphorylates a 47-kDa protein in control neutrophil cytoplasts but not in neutrophil cytoplasts from patients with the autosomal form of chronic granulomatous disease. *J. Biol. Chem.* 263: 2352-2357.
46. Okamura, N., S.E. Malawista, R.L. Roberts, H. Rosen, H.D. Ochs, B.M. Babior, and J.T. Curnutte. 1988. Phosphorylation of the oxidase-related 48K phosphoprotein family in the unusual autosomal cytochrome-negative and X-linked cytochrome-positive types of chronic granulomatous disease. *Blood* 72: 811-816.
47. Heyworth, P.G., C.F. Shrimpton, and A.W. Segal. 1989. Localization of the 47kDa phosphoprotein involved in the respiratory-burst NADPH oxidase of phagocytic cells. *Biochem. J.* 260: 243-248.
48. Clark, R.A., B.D. Volpp, K.G. Leidal, and W.M. Nauseef. 1990. Two cytosolic components of the human neutrophil respiratory burst oxidase translocate to the plasma membrane during cell activation. *J. Clin. Invest.* 85: 714-721.
49. Doussiere, J., F. Laporte, and P.V. Vignais. 1986. Photolabeling of a O_2-generating protein in bovine polymorphonuclear neutrophils by an arylazido $NADP^+$ analog. *Biochem. Biophys. Res. Commun.* 139: 85-93.
50. Umei, T., K. Takeshige, and S. Minakami. 1987. NADPH-binding component of the superoxide-generating oxidase in unstimulated neutrophils and the neutrophils from the patients with chronic granulomatous disease. *Biochem. J.* 243: 467-472.
51. Smith, R.M., J.T. Curnutte, and B.M. Babior. 1989. Affinity labeling of the cytosolic and membrane components of the respiratory burst oxidase by the 2',3'-dialdehyde derivative of NADPH. Evidence for a cytosolic location of the nucleotide-binding site in the resting cell. *J. Biol. Chem.* 264: 1958-1962.
52. Curnutte, J.T., and B.M. Babior. 1987. Chronic granulomatous disease. Adv.Hum.Genet.16: 229-297.
53. Clark, R.A., H.L. Malech, J.I. Gallin, H. Nunoi, B.D. Volpp, D.W. Pearson, W.M. Nauseef, and J.T. Curnutte. 1989. Genetic Variants of Chronic Granulomatous Disease: Prevalence of Deficiencies of Two Cytosolic Components of the NADPH Oxidase System. *N. Engl. J. Med.* 312: 647-652.
54. Weening, R.S., L. Corbeel, M. de Boer, R. Lutter, R. van Zwieten, and D. Roos. 1985. Cytochrome b deficiency in an autosomal form of chronic granulomatous disease: a third form of chronic granulomatous disease recognized by monocyte hybridization. *J. Clin. Invest.* 75: 915-920.
55. Wymann, M.P., V. von Tscharner, D.A. Deranleau, and M. Baggiolini. 1987. The onset of the respiratory burst in human neutrophils. Real-time studies of H_2O_2 formation reveal a rapid agonist-induced transduction process. *J. Biol. Chem.* 262: 12048-12053.
56. Dewald, B., T.G. Payne, and M. Baggiolini. 1984. Activation of NADPH oxidase of human neutrophils. Potentiation of chemotactic peptide by a diacylglycerol. *Biochem. Biophys. Res. Commun.* 125: 367-373.
57. Sklar, L.A., P.A. Hyslop, Z.G. Oades, G.M. Omann, A.J. Jesaitis, R.G. Painter, and C.G. Cochrane. 1985. Signal transduction and ligand-receptor dynamics in the human neutrophil. Transient responses and occupancy-response relations at the formyl peptide receptor. *J. Biol. Chem.* 260: 11461-11467.
58. Ohta, H., F. Okajima, and M. Ui. 1985. Inhibition by islet-activating protein of a chemotactic peptide-induced early breakdown of inositol phospholipids and Ca^{2+} mobilization in guinea pig neutrophils. *J. Biol. Chem.* 260: 15771-15780.
59. Okajima, F., T. Katada, and M. Ui. 1985. Coupling of the guanine nucleotide regulatory protein to chemotactic peptide receptors in neutrophil membranes and its uncoupling by islet-activating protein, pertussis toxin. A possible role of the toxin substrate in Ca^{2+}-mobilizing receptor-mediated signal transduction. *J. Biol. Chem.* 260: 6761-6768.
60. Curnutte, J.T., B.M. Babior, and M.L. Karnovsky. 1979. Fluoride-mdiated activation of the respiratory burst in human neutrophils. *J. Clin. Invest.* 63: 637-647.
61. Nasmith, P.E., G.B. Mills, and S. Grinstein. 1989. Guanine nucleotides induce tyrosine phosphorylation and activation of the respiratory burst in neutrophils. *Biochem. J.* 257: 893-897.
62. Buss, J.E., S.M. Mumby, P.J. Casey, A.G. Gilman, and B.M. Sefton. 1987. Myristoylated α subunits of guanine nucleotide-binding regulatory proteins. *Proc. Natl. Acad. Sci. USA* 84: 7493-7497.
63. Jones, T.L.Z., W.F. Simonds, J.J. Merendino Jr, M.R. Brann, and A.M. Spiegel. 1990. Myristoylation of an inhibitory GTP-binding protein α-subunit is essential for its membrane attachment. *Proc. Natl. Acad. Sci. USA* 87: 568-572.

64. Mumby, S.M., R.O. Heukeroth, J.I. Gordon, and A.G. Gilman. 1990. G-protein α-subunit expression, myristoylation, and membrane association in COS cells. *Proc. Natl. Acad. Sci. USA* **87**: 728-732.
65. Smith, C.D., C.C. Cox, and R. Snyderman. 1986. Receptor-coupled activation of phosphoinositide-specific phospholipase C by an N protein. *Science* **232**: 97-100.
66. Volpe, P., K.H. Krause, S. Hashimoto, F. Zorzato, T. Pozzan, J. Meldolesi, and D.P. Lew. 1988. "Calciosome," a cytoplasmic organelle: the inositol 1,4,5-trisphosphate-sensitive Ca^{2+} store of nonmuscle cells? *Proc. Natl. Acad. Sci. USA* **85**: 1091-1095.
67. Pozzan, T., D.P. Lew, C.B. Wollheim, and R.Y. Tsien. 1983. Is cytosolic ionized calcium regulating neutrophil activation? *Science* **221**: 1413-1415.
68. Billah, M.M., S. Eckel, T.J. Mullmann, R.W. Egan, and M.I. Siegel. 1989. Phosphatidylcholine hydrolysis by phospholipase D determines phosphatidate and diglyceride levels in chemotactic peptide-stimulated human neutrophils. Involvement of phosphatidate phosphohydrolase in signal transduction. *J. Biol. Chem.* **264**: 17069-17077.
69. von Tscharner, V., B. Prod hom, M. Baggiolini, and H. Reuter. 1986. Ion channels in human neutrophils activated by a rise in free cytosolic calcium concentration. *Nature* **324**: 369-372.
70. Grzeskowiak, M., V. Della Bianca, M.A. Cassatella, and F. Rossi. 1986. Complete dissociation between the activation of phosphoinositide turnover and of NADPH oxidase by formyl-methionyl-leucyl-phenylalanine in human neutrophils depleted of Ca^{2+} and primed by subthreshold doses of phorbol 12, myristate 13, acetate. *Biochem. Biophys. Res. Commun.* **135**: 785-794.
71. Dewald, B., M. Thelen, and M. Baggiolini. 1988. Two transduction sequences are necessary for neutrophil activation by receptor agonists. *J. Biol. Chem.* **263**: 16179-16184.
72. Repine, J.E., J.G. White, C.C. Clawson, and B.M. Holmes. 1974. The influence of phorbol myristate acetate on oxygen consumption by polymorphonuclear leukocytes. *J. Lab. Clin. Med.* **83**: 911-920.
73. Wilson, E., M.C. Olcott, R.M. Bell, A.H.Jr. Merrill, and J.D. Lambeth. 1986. Inhibition of the oxidative burst in human neutrophils by sphingoid long-chain bases. Role of protein kinase C in activation of the burst. *J. Biol. Chem.* **261**: 12616-12623.
74. Dewald, B., M. Thelen, M.P. Wymann, and M. Baggiolini. 1989. Staurosporine inhibits the respiratory burst and induces exocytosis in human neutrophils. *Biochem. J.* **264**: 879-884.
75. Baggiolini, M., B. Dewald, J. Schnyder, W. Ruch, P.H. Cooper, and T.G. Payne. 1987. Inhibition of the phagocytosis-induced respiratory burst by the fungal metabolite wortmannin and some analogues. *Exp. Cell Res.* **169**: 408-418.
76. Colditz, I., R. Zwahlen, B. Dewald, and M. Baggiolini. 1989. In vivo inflammatory activity of neutrophil-activating factor, a novel chemotactic peptide derived from human monocytes. *Am. J. Pathol.* **134**: 755-760.
77. Baggiolini, M., B. Dewald, and M. Thelen. 1988. Effects of PAF on neutrophils and mononuclear phagocytes. *Prog. Biochem. Pharmacol.* **22**: 90-105.
78. Wymann, M.P., P. Kernen, D.A. Deranleau, and M. Baggiolini. 1989. Respiratory Burst Oscillations in Human Neutrophils and Their Correlation with Fluctuations in Apparent Cell Shape. *J. Biol. Chem.* in press.
79. Wymann, M.P., P. Kernen, T. Bengtsson, T. Andersson, M. Baggiolini, and D.A. Deranleau. 1990. Corresponding oscillations in neutrophil shape and filamentous actin content. *J. Biol. Chem.* **265**: 619-622.

INDUCTION AND REGULATION OF INTERLEUKIN-8 GENE EXPRESSION

Robert M. Strieter,* Theodore Standiford,* Stephen W. Chensue, Keita Kasahara, and Steven L. Kunkel

Department of Pathology and *Internal Medicine
Division of Critical Care Medicine
University of Michigan Medical School
Ann Arbor, Michigan 48109-0602, USA

INTRODUCTION

A variety of disease states are characterized by a significant infiltrate of polymorphonuclear leukocytes. These acute disorders can range from immune complex-mediated inflammation, such as glomerulonephritis and rheumatoid arthritis, to more enigmatic diseases (e.g. adult respiratory distress syndrome and psoriasis). An interesting aspect of many acute inflammatory reactions is the infiltrate that occurs during the initial reaction is nearly a pure population of neutrophils. If the inflammatory reaction persists, the histological picture evolves into a heterogeneous population of immune cells. Therefore, the recruitment of cells from the vascular compartment to the inflammatory lesion is a dynamic process and changes as the immune response evolves. This characteristic of an inflammatory lesion is a good example of the host's ability to synthesize specific agents, which can elicit specific white blood cells, at specific time intervals of an inflammatory response.

INFLAMMATORY CELL ELICITATION

The sequence of events that result in leukocyte trafficking and recruitment is truly complex. In order for granulocytes to be the first peripheral white cell to be actively elicited, it must respond to a signal(s) in either a preferential or specific manner. Although the ability of the neutrophil to be the first cell to respond during inflammation has been of scientific interest for decades, the mechanism(s) that drives this response is not entirely known. A number of anomalies become apparent when comparing the movement of neutrophils with the recruitment of other inflammatory cell populations, such as monocytes. Both neutrophils and monocytes must initially bind to the endothelium via novel adherence proteins that are expressed on the surface of neutrophils and monocytes (CD11/CD18) and on the surface of the endothelial cells (ICAM and ELAM) (1,2). Both monocytes and neutrophils respond by moving to a number of common chemotactic factors, such as C5a and leukotriene B_4. Although the mechanisms that bind and move both types of inflammatory cells are shared, the arrival of neutrophils and monocytes to an inflammatory lesion is kinetically different. One obvious explanation is that a significant number of neutrophils are continuously marginated to the endothelial surface as a normal consequence of neutrophil physiology. Thus, the continuous

Chemotactic Cytokines, Edited by J. Westwick *et al.*
Plenum Press, New York, 1991

location of large numbers of neutrophils in juxtaposition to endothelial cells provides a source of neutrophils for rapid elicitation. On the contrary, monocytes do not appear to be marginated in great numbers and must interact with the endothelial cells during states of leukostasis and/or leukoconcentration. The temporal delay in physical contact between monocytes and endothelium may, in itself, be a reason for the delay in the arrival of monocytes to an inflamed area. A second reason for the apparent time difference in the recruitment of neutrophils and monocytes to an inflammatory lesion maybe the sheer difference in numbers of neutrophils as compared to monocytes. Since granulocytes comprise nearly 80% of the white blood cell pool, there may be more of these cells that can be elicited to an inflammatory lesion. Upon a histological assessment of an inflammatory cell infiltrate, the monocytes may actually be present in the lesion, but only in small numbers. Finally, a plausible explanation may be that specific factors are induced that selectively elicit neutrophils and not monocytes.

This last theory has recently received much attention, as a family of novel chemotactic cytokines have recently been isolated, cloned and expressed. One of these important factors has been termed neutrophil activating factor (3), granulocyte chemotactic factor (4), monocyte-derived chemotactic factor (5,6), neutrophil chemotactic factor (7) and interleukin-8 (IL-8) (8). This chemotactic cytokine has been shown to be highly chemotactic at nM concentrations (9), as well as cause neutrophils to undergo shape change and degranulation (9). The neutrophil activating events induced by IL-8 appear to occur via a calcium and protein kinase C dependent mechanism involving a GTP binding protein. This signal transduction mechanism of IL-8 is shared by a number of other chemotactic factors, including C5a and LTB_4.

These new chemotactic factors have been identified from a variety of cells and possess specificity with regard to their biological activity. The production of chemotactic factors which possess selectivity for the movement of certain inflammatory cells from the peripheral blood to the interstitium could be the major mechanism for the timely arrival of the appropriate inflammatory cell to an area of immunologic reactivity. Our laboratory has been interested in the selective recruitment of inflammatory cells via the induction and action of chemotactic cytokines. In the following review we present data describing the stimulus and cell specific induction of steady state IL-8 mRNA and examine factors which are important in the regulation of the expression of IL-8.

As noted above, one of the proximal events which lead to the movement of cells from the vessel lumen to the interstitium occurs via an active binding event. Yet this interaction must be reversible in order for the cells to move in response to a chemical signal. Our laboratory has shown that the involvement of IL-8 in this response is both stimulus and cell specific. In addition to mononuclear phagocytes (5,6,10), hepatoma cell lines (11), fibroblasts (12,13) and epithelial cells (14) can express steady state IL-8 mRNA when appropriately stimulated.

As shown in Table 1, lipopolysaccharide (LPS) was an efficacious trigger for the induction of IL-8 mRNA by human monocytes, alveolar macrophages and endothelial cells. On the contrary, LPS did not induce IL-8 mRNA expression by human fibroblasts, epithelial cells, or hepatoma cells. Interleukin-1α (IL-1α), IL-1β, and tumor necrosis factor-α (TNF) were potent stimuli for the induction of IL-8 by all of the cells studied. Cytokine specificity was demonstrated in this system as neither interleukin-6 (IL-6), interleukin-2 (IL-2) nor interleukin-4 (IL-4) could induce IL-8 mRNA expression. We have studied the expression of IL-8 by cytokine treated hepatoma cells and have shown that either IL-1α, IL-1β, or TNF could cause the production of significant IL-8 mRNA levels by 8 hours post challenge. At this time point the target cells have reached a 50% response. The production of IL-8 by stimulated hepatocytes did not reach a plateau level until 16-20 hours after activation. Northern blot analysis of steady state IL-8

Table 1. Stimulus specific and cellular source of IL-8 mRNA expression. Although LPS is a potent trigger for endothelial cell and macrophage IL-8 mRNA expression it is not effective for fibroblast or epithelial cells.

	Stimulus						
Cell Source	LPS	TNF	IL-1α	IL-1β	IL-2	IL-4	IL-6
Blood Monocyte	++++	+++	+++	++++	--	--	--
Alveolar Macrophage	++++	+++	+++	++++	--	--	--
Endothelial Cells	++++	+++	+++	++++	--	--	--
Fibroblast	--	++	+++	++++	--	--	--
Epithelial Cell	--	++	+++	++++	--	--	--

mRNA isolated from stimulated hepatoma cells demonstrated a similar scenario. As shown by laser densitometry (Figure 1), the cells challenged with IL-1 demonstrated a rapid rise in steady state IL-8 mRNA levels within the first 1-2 hours of exposure. Peak expression occurred at approximately 4 hours and persisted for the next 12 hours of the study. The induction of IL-8 mRNA by either TNF or IL-1 treated hepatocytes was shown to be concentration dependent, as TNF or IL-1 levels between 200 pg/ml and 20 ng/ml caused a dose related increase in steady state IL-8 mRNA. At concentrations above this level of cytokine, no further production of IL-8 mRNA could be expressed by the treated hepatoma lines.

One of the interesting aspects of steady state IL-8 mRNA expression by non-inflammatory cells is their inability to respond to an LPS challenge. This finding was true in the case of many primary isolates of epithelial cells and fibroblasts or epithelial and fibroblast cell lines. Human pulmonary fibroblasts given LPS over a wide dose range 100 pg - 10 μg/ml did not express of IL-8 mRNA or bioactivity (Figure 2). These same cells could biologically respond to TNF or IL-1 over a

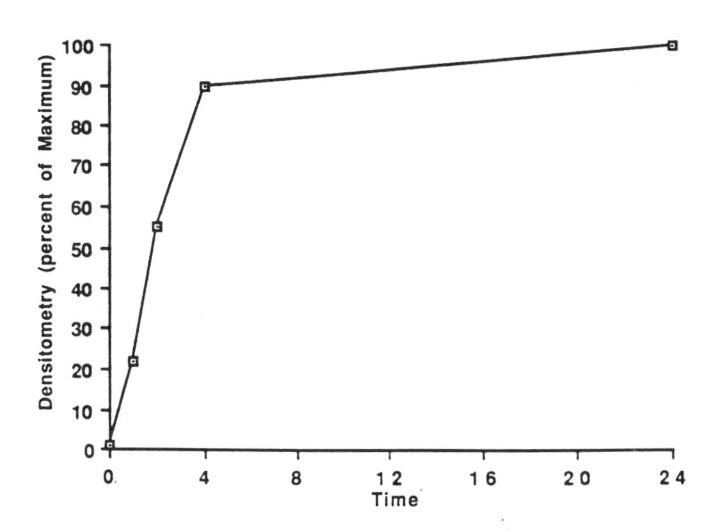

Figure 1. Laser densitometry of IL-8 mRNA expression by IL-1 treated hepatoma cells. Kinetic analysis demonstrated a rapid rise in IL-8 mRNA.

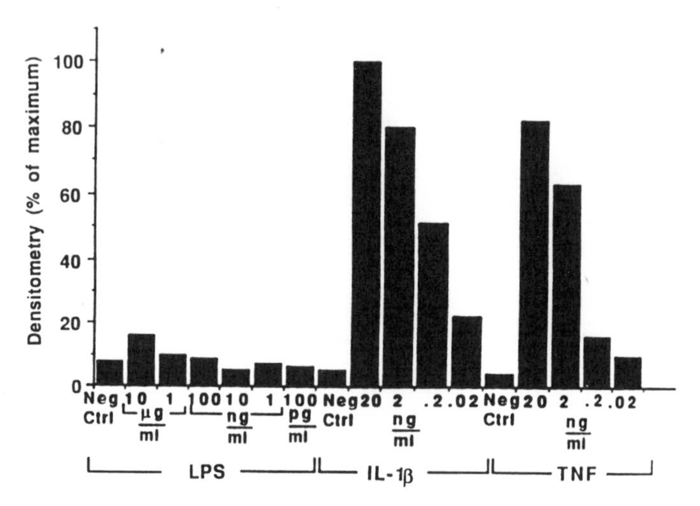

Figure 2. Laser densitometry of IL-8 mRNA expression by human fibroblasts treated with graded doses of LPS, IL-1b or TNF.

wide dose response range and express highly significant levels of IL-8 mRNA. These findings suggest that the non-immune cells must first wait for the host's inflammatory cell to generate stimulatory cytokines, such as IL-1 or TNF. Once this primary response has occurred a number of cells, in a cascade like fashion, can generate IL-8. This point is important as cells originally thought of as bystander cells or target cells of the inflammatory response must now be considered as true effector cells of inflammation. Also, the common location of epithelial cells and fibroblasts in all organ systems may prove useful as a cellular source of a potent neutrophil chemotactic factor.

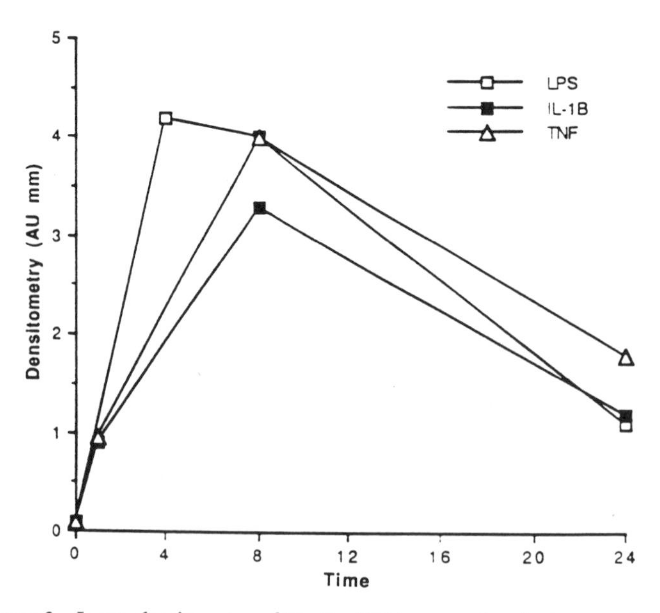

Figure 3. Laser densitometry of IL-8 mRNA expression by human alveolar macrophage treated with LPS, IL-1b or TNF.

The production of IL-8 by cells that comprise the alveolar architecture could provide important information regarding the rapid recruitment of granulocytes to the lung in such diverse diseases as ARDS (15), idiopathic pulmonary fibrosis (16), and asbestosis (17). To this end we have assessed the expression of steady state IL-8 mRNA by human alveolar macrophages recovered from normal volunteers. Alveolar macrophages ($2x10^6$ total) were exposed to either LPS (1 ug/ml), TNF (20 ng/ml) or IL-1β (20 ng/ml). At specific time intervals, total RNA was isolated from the alveolar macrophages and IL-8 mRNA was determined by Northern blot analysis. As shown by the laser densitometry scans in Figure 3, steady state IL-8 mRNA peaked between 4 and 8 hours post challenge and dropped to low, but detectable levels by 24 hours. In further studies, alveolar macrophages were challenged with graded doses of TNF and IL-8 mRNA expression was determined 4 hours later. TNF levels of 20 – 200 pg/ml were sufficient to induce the expression of IL-8 mRNA, which reached peak expression at 20 ng/ml. In all of the above studies, specific neutralizing antibody directed against human IL-8 could reduce a portion of the chemotactic activity. The degree of inhibition was dependent upon the stimulus with which the macrophages were challenged. The conditioned media from TNF, IL-1 and LPS treated human alveolar macrophages all demonstrated chemotactic activity for neutrophils and this activity could be suppressed by 47%, 44% and 31%, respectively, using anti IL-8 antibody. The ability of pulmonary fibroblasts, type II epithelial cells, and alveolar macrophages to generate a granulocyte chemotactic factor is highly significant in the context of lung inflammation, as granulocytes are often the predominant leukocyte found associated with many pulmonary disorders. The potential for pulmonary derived IL-8 to mediate many of these disorders may provide a new avenue to target therapeutic approaches.

REGULATION OF IL-*8 GENE EXPRESSION*

As previously discussed, the mechanism(s) by which inflammatory cells are recruited from the lumen of a vessel to an area of immunologic reactivity is poorly understood. Although the discovery of specific chemotactic cytokines has made inroads toward understanding cell movement, these agents have generated further questions of their own. For example, how is IL-8 gene expression regulated? To address this problem we have focused on the peripheral blood monocyte as the key cell for both the synthesis and regulation of IL-8. There is little doubt that LPS stimulated monocytes are a potent source of IL-8, as well as cytokines (IL-1 and TNF) that induce IL-8 expression by many other cells. Therefore, monocytes can either directly induce granulocyte migration via IL-8 synthesis or indirectly by secreting stimulatory levels of TNF or IL-1. This scenario places the monocyte at the head of a cascade designed to rapidly amplify the production of a single chemotactic factor. The pivotal position of this cell suggests that it must be highly susceptible to specific regulatory agents. In this regard we have identified a potent cytokine (IL-4) that can act as a suppressive compound for the regulation of monocyte-derived cytokines.

As shown in Figure 4, IL-4 is an extremely potent suppressive factor for the production of macrophage-derived TNF. This inhibitory effect is observed when IL-4 is added concomitantly with LPS. Other investigators have now shown that IL-4 can inhibit macrophage derived IL-1, GM-CSF, and PGE_2 (18). Our studies now demonstrate that IL-4 can both dose dependently, and in a time ordered sequence, inhibit monocyte IL-8 gene expression. This suppression of IL-8 mRNA by IL-4 was not stimulus dependent, as LPS, IL-1 and TNF-induced IL-8 mRNA was susceptible to regulation. As shown by the laser densitometry scans of the

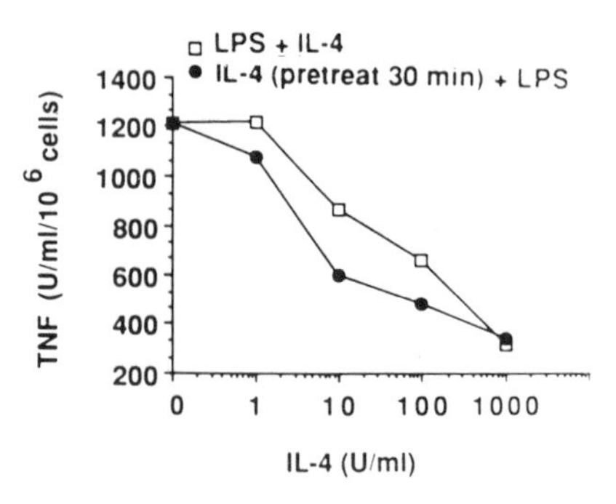

Figure 4. Suppression of TNF production by LPS treated activated murine macrophages in the presence of various doses of IL-4 added concomitantly with LPS (open box) or pretreated for 30 minutes prior to addition of LPS.

Northern blots in Figure 5, IL-4 over a broad dose response range (10 pg/ml–10 ng/ml) could regulate IL-8 steady state mRNA in response to either LPS or IL-1. Maximum suppression was found at 1 – 10 ng/ml, while significant suppression could be found at 10 pg/ml. Further preliminary studies suggest that the suppressive effects are dependent upon protein synthesis *de novo*, as the protein synthesis inhibitor, cycloheximide, abolished the IL-4 induced inhibitory effects on mRNA expression. This data has a number of exciting implications. IL-4 derived from local tissue or recruited lymphocytes may function *in vivo* by suppressing high levels of IL-8 and provide the "switch" mechanism that alters the inflammatory cell composition from one containing mostly neutrophils to one containing a more heterogeneous mononuclear cell component. An additional aspect, more important

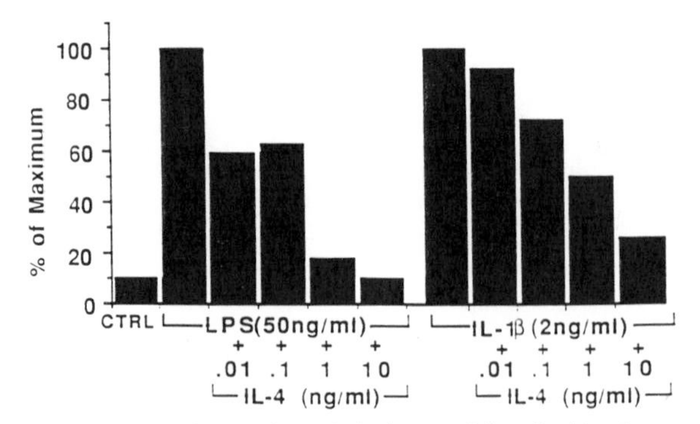

Figure 5. Effects of graded doses of interleukin-4 on monocyte IL-8 mRNA expression. Isolated monocytes were treated with either LPS or IL-1 and various doses of IL-4. The laser densitometry demonstrates the suppression of IL-8 mRNA levels by stimulated monocytes.

to global chronic inflammation, is that IL-4 may prove to regulate cell-mediated immune inflammation via its suppressive activity on monocyte-derived cytokines.

SUMMARY

Recent advances in understanding granulocyte elicitation have been made with the discovery and isolation of chemotactic cytokines. There is little doubt that these polypeptides will prove to be important mediators of disease process. Therefore, studies directed at understanding the production and regulation of interleukin-8 will continue to be a fertile area to explore mechanisms of disease processes and therapeutic targets.

Acknowledgements: The authors wish to thank Peggy Otto and Cassandra Narvab for expert secretarial support. This work was supported in part by NIH grants HL 31693, HL35276, HL-02401 and DK 38149. Steven L. Kunkel is an established investigator of the American Heart Association.

REFERENCES

1. Pohlman, T.H., K.A. Stanness, P.G. Beatty, H.D. Ochs, and J.M. Harlan. 1986. An endothelial cell surface factor(s) induced *in vitro* by lipopolysaccharide, interleukin-1, and tumor necrosis factor α increases neutrophil adherence by a CDW 18-dependent mechanism. *J. Immunol.* **136:** 4548-4559.
2. Schleimer, R.P. and B.K. Rutledge. 1985. Cultured human vascular endothelium acquire adhesiveness for neutrophils after stimulation with interleukin-1, endotoxin, and tumor promoting phorbol diesters. *J. Immunol.* **136:** 649-660.
3. Baggiolini, M., A. Walz and S.L. Kunkel. 1989. NAP-1/IL-8, a novel cytokine that activates neutrophils. *J. Clin. Invest.* **84:** 1045-1049.
4. Van Damme, J., J. Van Beeuman, A. Openakker, and A. Billau. 1988. A novel NH2-terminal sequence-characterized human monokine possessing neutrophil chemotactic, skin-reactivity and granulocytosis promoting activity. *J. Exp. Med.* **167:** 1364-1377.
5. Yoshimura,T., K. Matsushima, S. Tanaka, E.A. Robinson, E. Appella, J.J. Oppenheim, and E.J. Leonard. 1987. Purification of a human mononcyte-derived neutrophil chemotactic factor that shares sequence homology with other host defense cytokines. *Proc. Natl. Acad. Sci. USA* **84:** 9233-9245.
6. Matsushina, K., K. Marishita, T. Yoshimura, S. Lauw, Y. Kobayashi, W. Lew, E. Appella, N.F. King, E.J. Leonard, and J.J. Oppenheim. 1988. Molecular cloning of a human monocyte-derived neutrophil chemotactic factor (MDNCF) and the induction of MDNCF mRNA by interleukin-1 and tumor necrosis factor. *J. Exp. Med.* **167:** 1883-1896.
7. Strieter, R.M., S.L. Kunkel, J.J. Showell, D.G. Remick, S.H. Phan, P.A. Ward and R.M. Marks. 1989. Endothelial cell gene expression of a neutrophil chemotactic factor by TNF, IL-1, and LPS. *Science* **243:** 1467-1469.
8. Westwick, J., S.W. Li, and R.D. camp. 1989. Novel neutrophil-stimulating peptides. *Immunol. Today* **10:** 146-147.
9. Thelen, M., P. Peveri, P. Kermnen, V. VonTscharner, A. Walz, M. Baggiolini. 1988. Mechanisms of neutrophil activation by NAF, a novel monocyte-derived peptide agonist. *FASEB J.* **2:** 2702-2719.
10. Strieter, R.M., S.W. Chensue, M.A. Basha, T.J. Standiford, J.P. Lynch, M. Baggiolini, and S.L. Kunkel. 1990. Human alveolar macrophages gene expression of interleukin-8 by tumor necrois factor, lipopolysaccharide, and interleukin-1. *Am. J. Res. Cell Mol. Biol.* **2:** 321-326.
11. Thornton, A.J., R.M. Strieter, I. Lindley, M. Baggiolini, and S.L. Kunkel. 1990. Cytokine gene expression of a neutrophil chemotactic factor/IL-8 in human hepatocytes. *J.Immunol.* **144:** 2609-2613.
12. Strieter, R.M., S.H. Phan, H.J. Showell, D.G. Remick, J.P. Lynch, M. Genord, C. Raiford, M. Eskandari, R.M. Marks, and S.L. Kunkel. 1989. Monokine-induced neutrophil chemotactic factor gene expression in human fibroblasts. *J. Biol. Chem.* **264:** 10621-10626.

13. Mielke, V., J.G.J. Bauman, M. Sticherling, T. Ibs, A.G. Zomershoe, K. Seligmann, H. Henneicke, J.M. Schroder, W. Sterry, and E. Christophers. 1990. Detection of neutrophil-activating peptide NAP/IL-8 and NAP/IL-8 mRNA in human rIL-1 and rTNF stimulated human dermal fibroblasts. *J. Immunol.* **144**: 153-161.
14. Elner, V.M., R.M. Strieter, S.G. Elner, M. Baggiolini, I Lindley, and S.L. Kunkel. 1990. Neutrophil chemotactic factor (IL-8) gene expression by cytokine treated retinal pigmented epithelial cells. *Am. J. Pathol.* **136**: 745-750.
15. Parsons, P.E., A.A. Flower, T.M. Hyers, and P.M. Henson. 1985 Chemotactic activity in bronchoalveolar lavage fluid from patients with adult respiratory distress syndrome. *Am. Rev. Resp. Dis.* **132**: 490-493.
16. Hunnighake, G.W., J.E. Gradek, T.J. Lawley, R.G. Crystal. 1981. Mechanisms of neutrophil accumulation in the lungs of patients with idiopathic pulmonary fibrosis. *J. Clin. Invest.* **68**: 259-269, 1981.
17. Hayes, A.A., H. Rose, A.W. Musk, and B.W. Robinson. 1988. Neutrophil chemotactic factor release and neutrophil alveolitis in asbestos-exposed individuals. *Chest.* **94**: 521-525.
18. Hart, P.H., G.F. Vitti, D.R. Burgess, G.A. Whitty, D.S. Piccoli, and J.H. Hamilton. 1989. Potential anti-inflammatory effects of interleukin 4: Suppression of human monocyte tumor necrosis factor, interleukin-1 and prostaglandin E. *Proc. Natl. Acad. Sci. USA* **86**: 3803-3817.

REGULATION OF HUMAN INTERLEUKIN 8 GENE EXPRESSION AND BINDING OF SEVERAL OTHER MEMBERS OF THE INTERCRINE FAMILY TO RECEPTORS FOR INTERLEUKIN-8

Naofumi Mukaida, Atsushi Hishinuma, Claus O.C. Zachariae, Joost J. Oppenheim, and Kouji Matsushima

Biological Response Modifiers Program, Cancer Treatment Division
National Cancer Institute
Frederick, Maryland 21702-1201, USA

INTRODUCTION

Recruitment of leukocytes into tissues presumeably involves the local release of leukocyte activating and chemotactic mediators in response to injury-induced inflammatory reactions. We have recently purified major neutrophil, monocyte, and lymphocyte chemotactic factors to homogeneity from the conditioned media of human PBMC or monocytic cell line stimulated with mitogens and have shown that these chemotactic factors are structurally distinct from known cytokines, including IL-1 or TNF (1,2,3). Neutrophil chemotactic factor was structurally identical to T lymphocyte chemotactic factor and has been renamed IL-8 (3). IL-8 is distinct from the monocyte chemotactic and activating factor (MCAF) (2). cDNA cloning of IL-8 has been performed and the cDNA encodes a 99 amino acid precursor form of IL-8 (4). There is a typical signal peptide at the amino terminal end of the IL-8 precursor molecule and the mature processed soluble extracellular form of IL-8 consists of 72 amino acids. IL-8 shows considerable amino acid sequence homology to several human cytokines, such as β-thromboglobulin, platelet factor 4, gamma-IP 10 and GRO/MGSA. For ease of communication we will identify these cytokines as members of the intercrine family in this article. Such a generic term may provide useful in providing a name for members of this cytokine family based on their chromosomal location when they are precisely determined over the next few years. For example, cytokines such as IL-8 that are located on chromosome 4 can therefore be assigned a number in the intercrine α subfamily, whereas MCAF and the others on chromosome 17 can be assigned a number in the intercrine β subfamily.

Three dimensional structural analysis of recombinant IL-8 by both nuclear magnetic resonance (5) and X-ray crystallography (E. Baldwin et al., submitted for publication) has shown that IL-8 exists as a dimer. IL-8 dimer has a structural motif in which the top of a six-stranded anti-parallel β-sheet platform is derived from two, three-stranded Greek keys, one from each monomeric unit.

The monocyte chemotactic factor has also been purified (2) and molecularly cloned (6), and recombinant molecules have been expressed in CHO cells (K. Hirose et al., unpublished). Since this novel monocyte chemotactic factor also

Chemotactic Cytokines, Edited by J. Westwick *et al.*
Plenum Press, New York, 1991

induces superoxide production (7), lysosomal enzyme release (7), and increases the cytostatic activity of monocytes against several types of human tumor cell lines in vitro (2), we termed this factor monocyte chemotactic and activating factor (MCAF). MCAF mRNA encodes a 99 amino acid precursor MCAF.

There is a typical signal peptide at the amino terminal end of the pre-MCAF molecule and the mature extracellular form of MCAF consists of 76 amino acids. MCAF shows considerable (30 to 40%) amino acid sequence homology with a number of other members of the intercrine β subfamily whose genes are located on the chromosome 17 region in man and are also characterized by having two of their 4 cysteines in adjoining locations. The other intercrine β members are JE, LD 78, RANTES, TCA-3 and ACT-2. MCAF also shows significant amino acid sequence homology with IL-8 (21%). Purified unlabelled MCAF competes with ^{125}I-labelled MCAF in binding to human PBMC, whereas the neutrophil chemo-attractant, IL-8 did not, suggesting the existence of specific receptors for MCAF on human PBMC which are distinct from receptors for IL-8 (7).

In this chapter, we will describe the production of IL-8 by various cell types, the molecular analysis of the regulation of IL-8 gene expression by IL-1, TNF, or PMA, and the identification of several members of the intercrine family that compete with IL-8 for binding to its receptor on neutrophils.

PRODUCTION OF IL-8 BY VARIOUS CELL TYPES

IL-8 was originally purified from conditioned media of LPS-stimulated human PBMC (1) and the predominant producing cells were determined to be monocytes (8). The availability of cDNA for IL-8 facilitated identification of other producing cells and their stimulants. As shown in Table 1, many types of somatic cells have been shown to produce IL-8 by stimulation with IL-1 and TNF as well as by various types of mitogens, lectins, and tumor promoters. On the other hand, less information is available on suppressors of the production of IL-8. Glucocorticoids have been shown to suppress the production of IL-8 by LPS-stimulated human PBMC (9). TGFβ (unpublished observation) and interleukin 4 also reduce production of monocyte-derived IL-8 (10). In this meeting, a 5'-lipoxygenase inhibitor and 1,25-$(OH)_2$-Vitamin D_3 were also shown to suppress the production of IL-8 in vitro (11).

Table 1. IL-8-producing cells

Cell types	Stimulants
monocyte/macrophages	IL-1, TNF, LPS, poly I:C, uric acid crystals, PMA, plant polysaccharide etc.
T lymphocytes	PMA + calcium ionophore
Dermal fibroblasts	IL-1, TNF, PMA, poly I:C
Keratinocytes	IL-1, TNF + rIFN, LPS
Vascular endothelial cells	IL-1, TNF, PMA, LPS
Melanocytes	IL-1, TNF
Various types of tumor cell lines	many of them stimulated with IL-1, TNF or PMA

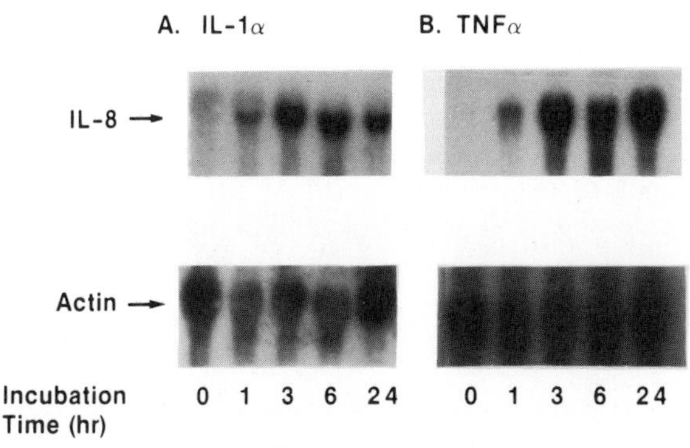

Figure 1. Kinetics of IL-8 mRNA accumulation in a human fibrosarcoma cell line, 8387, when stimulated with IL-1α or TNFα.

ACTIVATION OF THE IL-8 GENE AFTER STIMULATION BY IL-1, TNF, OR PMA

Cloning of IL-8 cDNA enabled us to analyze how mRNA for this gene was expressed in a various cell types including human PBMC, dermal fibroblasts, keratinocytes, endothelial cells and hepatocytes. Among cytokines tested, only IL-1 and TNF could rapidly and dramatically induce IL-8 mRNA in the above-mentioned types of cells. As shown in Figure 1, this increase could be observed within 1 h after stimulation with IL-1 or TNF and the expression reached a maximal level within 3 h after stimulation, gradually decreasing thereafter. When cells were stimulated with PMA, a similar pattern of mRNA expression was observed (data not shown).

Cycloheximide enhanced the mRNA expression induced by IL-1, TNF, and PMA (Figure 2). This enhancement of mRNA expression was also seen by 1 hr after stimulation. These results indicate that the induction of IL-8 mRNA occurred

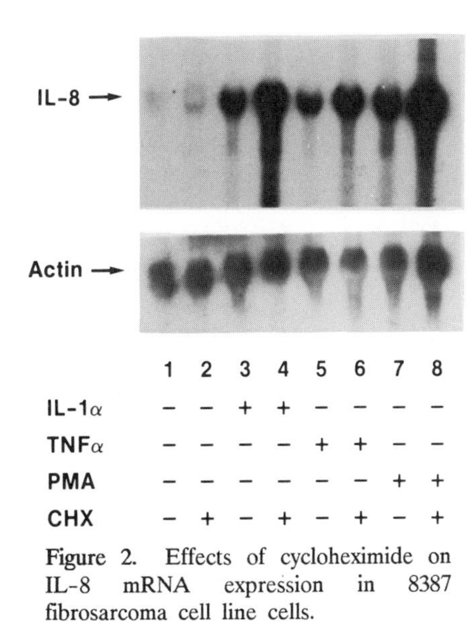

Figure 2. Effects of cycloheximide on IL-8 mRNA expression in 8387 fibrosarcoma cell line cells.

in the absence of new protein synthesis and that synthesis of a repressor protein, which can be blocked by cycloheximide, may be induced to regulate mRNA expression negatively. Run-off assays performed demonstrated that this mRNA induction is at least partly based on the activation of transcription in various cells including a human myelomonocytic cell line, HL-60, human endothelial cells, and a human fibrosarcoma cell line, 8387. In HL-60 cells, the mRNA stability was increased by stimulation with LPS or PMA. At present, however, it remains to be determined whether IL-1 or TNF can also increase the stability of IL-8 mRNA in other cell types.

DETERMINATION OF IL-1, TNF, AND PMA RESPONSIVE ELEMENT IN THE 5'-FLANKING ENHANCER REGION OF THE IL-8 GENE

IL-1, TNF and PMA activated the IL-8 gene directly at the transcriptional level in the absence of new protein synthesis. Therefore, we assumed that IL-8 gene activation provides a good model for studying the mechanism of regulation of gene expression by these three agents at the molecular level. In order to elucidate the mechanism, we first cloned and determined the entire sequence of the genomic IL-8 gene including the 1.5 kb 5'-flanking region (9). IL-8 genomic DNA consists of 4 exons and 3 introns with a single "TATA"- and "CAT"- like structure. The 5'- flanking region of the IL-8 gene shows no overall sequence similarity

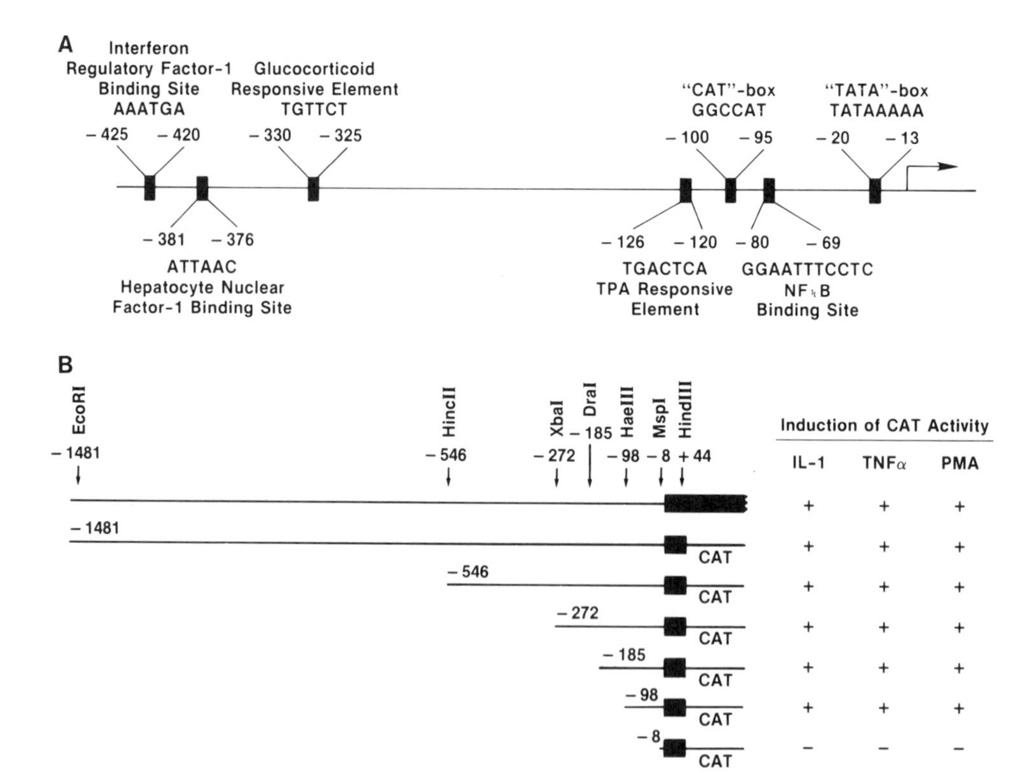

Figure 3A. Potential binding site for known nuclear factors.

Figure 3B. Results of CAT assay using several 5'- deletion mutants which are linked with a CAT expression vector.

with other cytokines or acute phase reactant genes, whose expression is also induced by IL-1 or TNF. However, the 5'- flanking region of the IL-8 gene contains several potential binding sites for known nuclear factors including AP-1, AP-2, NFkB, hepatocyte nuclear factor-1, glucocorticoid receptor and interferon regulatory factor-1 (Figure 3A).

In order to examine whether the 5'- flanking region of the IL-8 gene contains responsive element for stimulation by IL-1, TNF and PMA, sequentially deleted 5'- flanking DNA fragments were ligated into a PUX-CAT (chloramphenicol acetyl transferase) expression vector and were transfected into a human fibrosarcoma cell line, 8387, by calcium phosphate co-precipitation. 14 h after the transfection, cells were stimulated with IL-1, TNF or PMA. The region from -98 to -9 bp contained the functional responsive element required for responsiveness to IL-1, TNF and PMA (Figure. 3B).

Several additional deletion mutants, as shown in Figure 4A, were ligated into a PUX-CAT and were transfected into 8387 cells as described above. The results indicated that the sequence from -94 to -71 bp is minimally sufficient for conferring responsiveness to IL-1, TNF and PMA. This region shows considerable

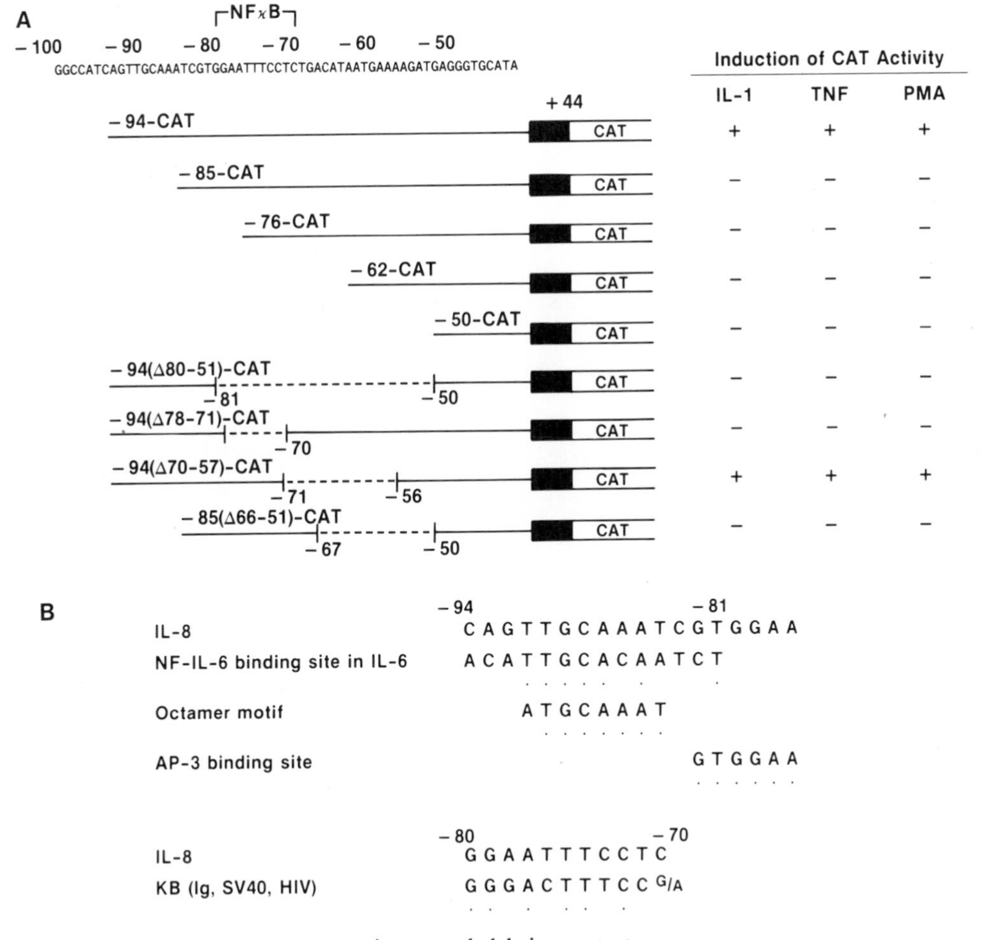

Figure 4A. Results of CAT assays using several deletion mutants.

Figure 4B. Comparison of the sequences between -94 and -71 bp with potential binding site for nuclear factors.

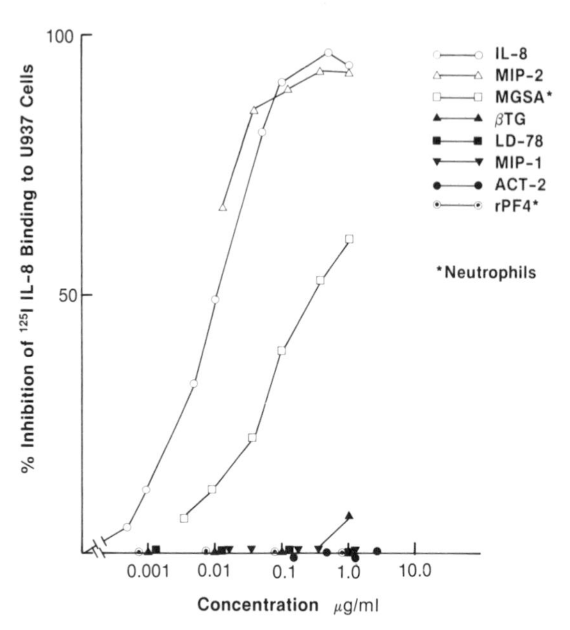

Figure 5. Competition of members of intercrine family for binding site of IL-8 on human neutrophils.

similarity to the potential binding sites for nuclear factors including NFkB, C/EBP-like factor, NF-IL 6, AP-3 and octamer motif binding proteins (Fig.4B). CAT assays using point-mutated CAT vector revealed that IL-8 gene expression required the cooperation of two distinct *cis*-elements; one being the potential binding site for NFkB-like factor and the other for a C/EBP-like factor. Identification of the nuclear factors which bind to this region is now in progress.

GRO/MGSA AND MIP2 SHARE THE RECEPTOR FOR IL-8

In an effort to determine whether or not various members of the intercrine family utilize the same cell surface receptors, we tested those intercrine members that were made available to us for their capacity to compete with the binding of radiolabelled IL-8 to receptors on human peripheral blood neutrophils or U937 cells. We observed that muMIP-2 (kindly provided by Drs. Barbara Sherry and Tony Cerami) binds equally as well as IL-8 to receptors for IL-8 on human neutrophils (Figure 5). Furthermore, recombinant huGRO/MGSA (kindly provided by Dr. Ann Richmond) binds almost as well as IL-8 to receptors for IL-8. The affinity of binding of MGSA was only 2 to 3 fold lower than that of IL-8 for neutrophil binding sites (Figure 5). On the other hand, the other available intercrines including huMCAF, rhuACT-2 (from Drs. Warren Leonard and Ulrich Siebenlist), rhuLD78 (from Drs. Toshio Hattori and Ulrich Siebenlist), muMIP-1 (from Drs. Barbara Sherry and Tony Cerami), platelet factor 4 and the carboxyl-terminal 13 amino acids of PF4 (from Dr. Jeanette Thorbecke), and CTAPIII (from Dr. Brandt) did not compete at all with the binding of IL-8. NAPII which is a breakdown product of CTAPIII and reported to be a potent neutrophil chemoattractant (12), interfered with IL-8 binding to only a limited extent and was

1000 fold less potent in binding to IL-8 receptors than IL-8 in our experiments. Furthermore, the preparation of NAPII provided to us by Dr. Brandt also failed to chemoattract neutrophils.

These observations suggest that murine MIP2 and GRO/MGSA are functionally related to IL-8. Indeed all 3 factors are documented to be chemoattractants and activators of neutrophils (1,13,14). Whether the IL-8 receptor is the only receptor utilized by these 3 factors on neutrophils and other somatic cells remains to be established. There are reports in this volume that IL-8, like MGSA acts on (chemoattracts) melanoma cells (15) and that IL-8, like MIP-2 may have some activating effects on monocytes, e.g. promoting adhesion of monocytes to endothelial cells (16). On the other hand, some members of the intercrine family such as PF4, the C13 end of PF4, and NAPII are reported to have neutrophil chemoattractant activity, although these observations are controversial. When more of these intercrine members become available, the pattern of receptor utilization should become clearer.

SUMMARY AND CONCLUSION

IL-8 is produced by a wide variety of cells in response to polyclonal mitogens and cytokines. Northern blotting analysis revealed that IL-1, TNF and PMA could induce rapid expression of IL-8 mRNA in the absence of new protein synthesis. Nuclear run-off assays using different cell types demonstrated that IL-8 mRNA expression could at least be partly due to the activation of transcription.

Cloning and determination of the entire sequence of IL-8 genomic DNA enabled us to explore the functional significance of the 5'- flanking enhancer region of the IL-8 gene by employing CAT assays. The results indicated that the region spanning from -94 to -71 bp is minimally sufficient for conferring responsiveness to IL-1, TNF and PMA. Further analysis using point-mutations revealed that this region consisted of two distinct *cis*-elements; one being the potential binding site for NFkB-like and the other for a C/EBP-like factor. These results suggested that all three stimuli, IL-1/TNF/PMA, modulate the identical combination of nuclear factors possibly by phosphorylation.

We previously reported that these three stimuli activated the same serine protein kinase which phosphorylates identical 65 kDa and 74 kDa cytosol proteins in human PBMC (17). This IL-1/TNF/PMA-activated protein kinase is distinct from protein kinase A, protein kinase C or casein kinase in substrate specificity; in Ca and phospholipid dependency; in cyclic nucleotide dependency; and sensitivity to protein kinase inhibitors. Taken collectively, IL-1/TNF/PMA may activate a common serine protein kinase and this protein kinase may in turn directly or indirectly modulate several nuclear factors.

Modulation of these nuclear factors may lead to the augmentation of binding of these nuclear factors with IL-1/TNF/PMA responsive *cis*-elements of the genes, thus enhancing the transcription of the genes. Purification of this IL-1/TNF/PMA activated protein kinase as well as identification of the nuclear factors which bind to the IL-1/TNF/PMA responsive elements will clarify the intracellular signals utilized by these stimuli at the molecular level.

Binding competition studies revealed that MIP2 and GRO/MGSA exhibit high binding affinity for IL-8 receptor sites on neutrophils. NAPII competed 1000 fold less well for binding to IL-8 receptors than IL-8. In contrast, MIP-1, ACT-2, LD78, MCAF, PF4 and the carboxy terminal 13 amino acids of PF4 and CTAPIII did not bind at all to IL-8 receptors. Thus, MIP-2, GRO and IL-8 may be functionally closely related.

REFERENCES

1. Yoshimura, T., K. Matsushima, S. Tanaka, E.A. Robinson, E. Appella, J.J. Oppenheim, and E.J. Leonard. 1987. Purification of human monocyte-derived neutrophil chemotactic factor that shares sequence homology with other host defense cytokines. *Proc. Natl. Acad. Sci. USA* **84**: 9233-9237.
2. Matsushima, K., C.G. Larsen, G.C. Dubois, and J.J. Oppenheim. 1989. Purification and characterization of a novel monocyte chemotactic and activating factor produced by a human myelomonocytic cell line. *J. Exp. Med.* 169: 1485-1490.
3. Larsen, C.G., A.O. Anderson, E. Appella, J.J. Oppenheim, and K. Matsushima. 1989. Identity of chemotactic cytokine for T-lymphocytes with neutrophil activating protein (NAP-1): A candidate interleukin 8. *Science* 243: 1464-1466.
4. Matsushima, K., K. Morishita, T. Yoshimura, S. Lavu, Y. Kobayashi, W. Lew, E. Appella, H.F. Kung, E.J. Leonard, and J.J. Oppenheim. 1988. Molecular cloning of a human monocyte derived neutrophil chemotactic factor (MDNCF) and the induction of MDNCF mRNA by interleukin 1 and tumor necrosis factor. *J. Exp. Med.* 167: 1883-1893.
5. Clore, G.M., E. Appella, M. Yamada, K. Matsushima, and A.M. Gronenborn. 1990. The three-dimensional structure of interleukin 8 in solution. *Biochemistry* 29: 1689-1696.
6. Furutani, Y., H. Nomura, M. Notake, Y. Oyamada, T. Fukui, M. Yamada, C.G. Larsen, J.J. Oppenheim, and K. Matsushima 1989. Cloning and sequencing of the cDNA for human monocyte chemotactic and activating factor (MCAF). *Biochem. Biophys. Res. Commun.* 159: 249-255.
7. Zachariae, C.O.C., A.O. Anderson, H.L. Thompson, E. Appella, A. Mantovani, J.J. Oppenheim, and K. Matsushima. 1990. Properties of monocyte chemotactic and activating factor (MCAF) purified from human fibrosarcoma cell line. *J. Exp. Med.* 171: 2177-2182.
8. Yoshimura, T., K. Matsushima, J.J. Oppenheim, and E.J. Leonard. 1987. Neutrophil chemotactic factor produced by lipopolysaccharide stimulated human blood mononuclear leukocytes. I. Partial characterization and separation from interleukin 1. *J. Immunol.* 139: 788-793.
9. Mukaida, N., M. Shiroo, and K. Matsushima. 1989. Genomic structure of the human monocyte-derived neutrophil chemotactic factor (MDNCF), interleukin 8. *J. Immunol.* 143: 1366-1371.
10. Standiford, T. J., R. M. Strieter, S. W. Chensue, J. Westwick, K. Kasahara, and S. L. Kunkel. 1990. IL 4 inhibits the expression of IL-8 from stimulated human monocytes. In J. Westwick et al. (Eds). *Chemotactic cytokines: biology of the inflammatory peptide supergene family*, (in press).
11. Larsen, C. G., M. Kristensen, K. Paludan, B. Deleuran, M. K. Thompsen, K. Kragballe, and K. Matsushima. 1990. Endogenous and synthetic inhibitors of the production of chemotactic cytokines. In J. Westwick et al. (Eds). *Chemotactic cytokines: biology of the inflammatory peptide supergene family* (in press).
12. Walz, A. and M. Baggiolini. 1990. Generation of the neutrophil activating peptide NAP-2 from platelet basic protein or connective tissue-activating peptide III through monocyte proteases. *J. Exp. Med.* 171: 449-454.
13. Schroeder, J-. M., N. L. M. Pearson, and E. Christophers. 1990. Lipopolysaccharide-stimulated human monocytes secrete apart from NAP-1/IL-8, a second neutrophil-activating protein. *J. Exp. Med.* 171: 1091-1100.
14. Wolpe, S. D., G. Davatelis, B. Sherry, B. Beutler, D. Hesse, H. T. Nguyen, L. L. Moldawer, C. F. Nathan, S. F. Lowry and A. Cerami. 1988. Macrophages secrete a novel heparin-binding protein with inflmmatory and neutrophil chemotactic properties. *J. Exp. Med.* 167: 570-581.
15. Wang, J. M., G. Taraboletti, K. Matsushima, J. Van Damme, and A. Mantovani. Induction of haptotactic migration of melanoma cells by NAP 1/IL-8. In J. Westwick et al. (Eds). *Chemotactic cytokines: biology of the inflammatory peptide supergene family*, (in press).
16. Brown, K. A., F. Le Roy, G. Noble, K. Bacon, R. Camp, A. Vora, and D. C. Dumonde. 1990. IL-8 acts on monocytes to increase their attachment to cultured endothelium and enhance their expression of surface adhesion molecules. In J. Westwick et al. (Eds). *2nd Int. Symp. on Chemotactic Cytokines*, (in press).
17. Shiroo, M. and K. Matsushima. 1990. TNF and IL-1 phosphorylate identical cytosolic 65 and 74 kDa proteins in human peripheral blood mononuclear cells. *Cytokine* 2: 13-20.

FORMATION AND BIOLOGICAL PROPERTIES OF NEUTROPHIL ACTIVATING PEPTIDE 2 (NAP-2)

Alfred Walz,[1] Roland Zwahlen,[2] and Marco Baggiolini[1]

[1]Theodor-Kocher Institute and
[2]Institute for Veterinary Pathology
University of Bern
CH-3000 Bern 9, Switzerland

INTRODUCTION

Neutrophil-activating peptide 2 (NAP-2) was isolated from stimulated cultures of human blood mononuclear cells (1). NAP-2 consists of 70 amino acids and is structurally related to NAP-1/IL-8 (2), a peptide produced by a variety of cells upon induction with interleukin-1 or tumor necrosis factor alpha (3), and to melanoma growth-stimulatory activity (MGSA) which was shown to be mitogenic for cultured human melanoma cells (4, 5). NAP-1/IL-8 and MGSA are potent chemotactic agents for human neutrophils *in vitro* and *in vivo* (6, 7), and induce cytosolic free calcium changes, the respiratory burst and exocytosis (8). The amino acid sequence of NAP-2 corresponds to part of the sequence of platelet basic protein (PBP) (9) and its derivative, connective tissue activating peptide III (CTAP-III, ref 10) and to other inactive cleavage products which were recently identified (11). NAP-2 also shows structural homology to platelet factor 4 (PF-4), another peptide contained in the platelet alpha granules (12).

NAP-2, which is not found in the α-granules of platelets, can be obtained by proteolytic cleavage from its precursors PBP and CTAP-III in the presence of monocytes or monocyte supernatants (13), suggesting that a common precursor is differentially processed into a number of related peptides with varying biological activities. Cloning of a complete cDNA for PBP and subsequent northern analysis identified a single species of mRNA in megakaryocytes, confirming the presence of a common precursor peptide (14). CTAP-III was reported to stimulate synthesis of DNA, hyaluronic acid, sulfated glycosaminoglycan chains and proteoglycan protein in human synovial fibroblast cultures. These activities were not observed with PBP, its 9 amino acid longer precursor (15).

We have compared the effects of highly purified NAP-2, CTAP-III, PBP, PF4, MGSA and NAP-1/IL-8 on human neutrophils. NAP-1/IL-8, MGSA and NAP-2 are similar in potency to activate chemotaxis and cytosolic free calcium rise, but vary in their ability to induce exocytosis and the respiratory burst response. The other peptides tested were practically inactive (16).

Chemotactic Cytokines, Edited by J. Westwick *et al.*
Plenum Press, New York, 1991

MATERIALS AND METHODS

Biological assays

Neutrophil chemotaxis (17), cytosolic free calcium changes (8) and elastase release (8) were determined according to established methods.

In vivo activity

Three rabbits (2 male, 1 female) were used for the study. 100 ul of the chemoattractants in pyrogen free saline were injected intradermally using 26-gauge hypodermic needles. Two sites per animal and dose (total 6 sites per dose) were injected. Four hours after injection skin samples were excised and fixed in 4% buffered formaldehyde. Full thickness skin biopsies were taken from each site, embedded in paraffin, routinely processed and stained with hematoxylin and eosin. Neutrophil infiltration in the skin was evaluated semiquantitatively using a scoring method. Coded samples were scored for the extent of extravascular neutrophil accumulation separately in the upper dermis (reaching from epithelium to the deepest adnexal structures) and in the lower dermis (from the deepest adnexal structures to the panniculus carnosus). Scores were as follows: 0, less than 6 neutrophils; 1, 6 to 40; 2, more than 40 neutrophils with moderate number of focal collections and/or relatively few scattered neutrophils; 3, more than 40 neutrophils forming extensive foci and relative few scattered neutrophils or relative few foci and moderate numbers scattered; 4, extensive foci and a marked number of neutrophils scattered. The scores for upper and lower dermis were added to yield a total score (range 0 to 8).

NAP-2 FORMATION

NAP-2 was originally found together with NAP-1/IL-8 in cultures of human blood mononuclear cells stimulated with LPS or lectins (1). Subsequently it was demonstrated that the formation of NAP-2 depends on the presence of platelets in the monocyte cultures (18). Neither platelets nor monocytes alone produce NAP-2. When monocytes were stimulated with LPS or phytohemagglutinin NAP-1/IL-8 was obtained in addition and accounted for the bulk of the neutrophil-stimulating activity. Monocyte stimulation did not markedly influence the yield of NAP-2. But, under these conditions, three amino-terminal NAP-2 variants, consisting of 73, 74 and 75 residues were also found (Figure 1). No NAP-1/IL-8 or NAP-2 was detected when lymphocytes were used instead of monocytes. NAP-2 and the three amino-terminal variants were also produced when the release supernatant of thrombin-stimulated platelets was added to cultures of resting and stimulated monocytes (13). Furthermore, NAP-2 and its variants were obtained in a cell-free system by incubating purified CTAP-III with the conditioned medium from monocyte cultures (13). A NAP-2 variant, consisting of 72 residues was recently identified in preparations of CTAP-III that had been stored for several years (15). This variant was not purified to homogeneity and was not characterized in terms of its biological activities on human neutrophils. However, on the basis of our results with NAP-2 variants 73, 74, and 75, we assume that it will also be biologically active. A NAP-2-like peptide was isolated from CTAP-III cleaved with porcine elastase and bovine chymotrypsin, however, no activity was demonstrated. Upon cleavage of CTAP-III with human elastase no neutrophil-stimulating activity was released (Car, Walz, unpublished observations).

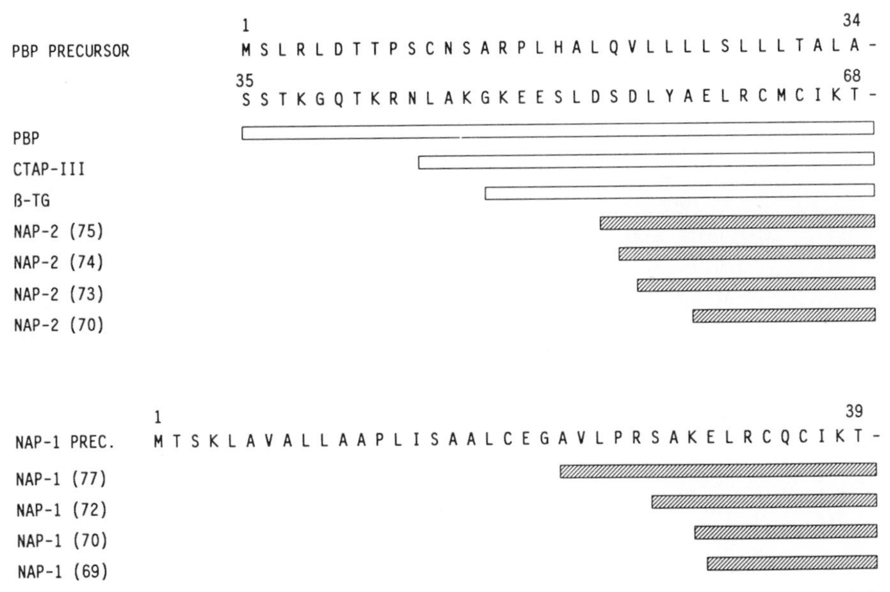

Figure 1. Amino-terminal sequences of NAP-2 and NAP-1/IL-8 variants and their precursors. Hatched bars indicate peptides with neutrophil-stimulating activity.

GENE CLONING

The cDNA coding for precursor sequence of PBP, CTAP-III, β-thromboglobulin and NAP-2 has been cloned from a lamda gt11 expression library which was prepared using mRNA isolated from human platelets (14). The open reading frame of the clone codes for a 128 amino acid peptide. 94 amino acids are identical to the protein sequence reported for human PBP (Figure 1). The 35 amino acid leader sequence as well as the coding region for PBP show significant homology to the sequence of PF4, another peptide contained in the α-granules of platelets and to GRO/MGSA. Interestingly there is no homology between the leader sequences of PBP and NAP-1/IL-8, possibly due to different secretory mechanisms for the two types of peptides. PBP and PF4 are targeted to the α-granules, whereas NAP-1/IL-8 is secreted. Northern blot analysis with platelet and megakaryocyte mRNA showed only one species of mRNA (0.8 kb), whereas mRNA from hepatocytes and from HEL cells (megakaryocytic cell line) were negative. A synthetic gene for NAP-2 has been constructed and was successfully expressed in E. coli (Walz, unpublished results).

Table 1. Neutrophil responses to NAP-1/IL-8, NAP-2 and GRO/MGSA.

	NAP-1	NAP-2	GRO/MGSA
Ca^{2+} rise	+++	+++	+++
Chemotaxis	+++	+++	+++
Exocytosis	+++	++	++
Respiratory burst	++	+	+
PMN infiltration	+++	+++	+++

BIOLOGICAL ACTIVITIES *IN VITRO*

NAP-2 was found to be a powerful activator of human neutrophils *in vitro*, inducing cytosolic free calcium changes, chemotaxis and exocytosis at concentrations between 0.3 and 10 nM. By contrast, its precursors PBP and CTAP-III had little, if any, activity at concentrations up to 100 nM (16). Highly purified PF4, which was reported earlier to be chemotactic for neutrophils, monocytes and fibroblasts and to induce neutrophil granule release (19), was at least 1000-fold less effective than NAP-2 on neutrophil stimulation. NAP-2 was equipotent with NAP-1/IL-8 and GRO/MGSA as a stimulus of cytosolic free calcium changes and chemotaxis, but weaker than these and other agonists as a stimulus of exocytosis (Table 1, ref 6, 16). NAP-2 and GRO/MGSA are inducers of the respiratory burst in human neutrophils. Their efficiency, however, is low compared to the activity of NAP-1/IL-8 or other agonists such as f-Met-Leu-Phe or C5a (20). N-amino-terminal variants of NAP-2 containing 75, 74, and 73 amino acids were found to induce cytosolic free calcium changes and exocytosis, but their biological activity was about 10 times lower than that of NAP-2. Amino-terminal variants were also isolated for NAP-1/IL-8 (Figure 1, ref. 21).

IN VIVO ACTIVITY

In vivo activity of NAP-2 was tested in rabbits by intradermal injections of 10^{-9} to 10^{-12} mol/site. Similar concentrations of NAP-1/IL-8 were used for comparison. Neutrophil infiltrates within the skin were scored in the lower and upper dermis 4 hours after injection. Total scores for the skin samples were obtained by adding both counts. The scores for NAP-1/IL-8 and NAP-2 were essentially identical at all 4 concentrations tested (Figure 2). A gradual decrease of neutrophil scores was recorded with decreasing concentrations of chemoattractants, reaching background levels at 10^{-11} to 10^{-12} mol/site. At 10^{-9} and 10^{-10} mol/site a massive neutrophil infiltration from venules into the surrounding tissue was observed (Figure 3 b, c). No neutrophil emigration occured from arterioles (Figure 3 c, upper part). As seen in Figure 3 c (lower part) neutrophils appear to

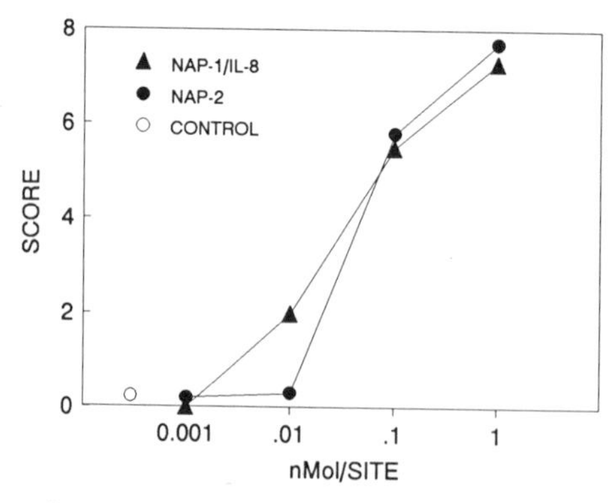

Figure 2. Neutrophil accumulation in rabbit skin 4 hours after injection of NAP-2 (10^{-12} to 10^{-9} mol/site). Each point represents score values from 6 injection sites (2 sites per animal). Scores were determined from neutrophil counts in the upper dermis and in the lower dermis (range 0-8, details in Materials and Methods).

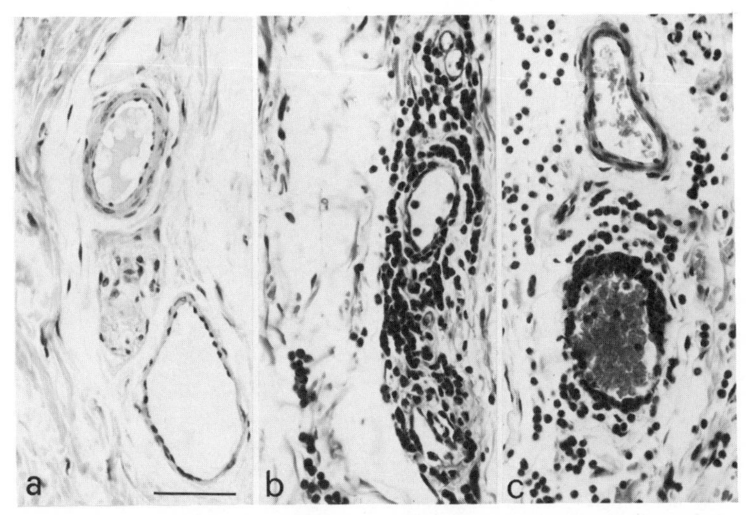

Figure 3. Histology of rabbit lower dermis 4 hours after intradermal injection of (a) pyrogen-free saline control; (b) 10^{-9} mol NAP-1/IL-8; (c) 10^{-9} mol NAP-2. Bar: 50 μm.

accumulate in the venules below the endothelial cell layer along the basal membrane before crossing the vessel wall. The same effect was observed with NAP-1, but not with other stimuli. 24-48 hrs after injection of the chemotactic peptides the perivascular tissue had a normal appearance and most of the neutrophils had disappeared. No monocytes or lymphocytes were seen. The absence of perivascular damage suggests that secretion of proteases from neutrophils is minimal.

STABILITY IN SERUM

An important question concerning the biological role of NAP-2 in the circulation is its stability in serum. NAP-2 was compared with C5a, which is rapidly deactivated by serum (Figure 4A). No neutrophil-activating activity was observed in the absence of C5a (Figure 4B). NAP-2 (10^{-6}M) was incubated in four different human sera at 37°C and after 25, 60 and 120 minutes activity was determined by

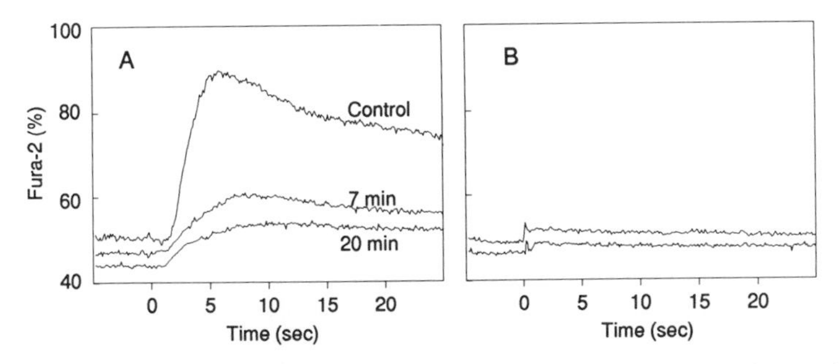

Figure 4. Stability of C5a in human serum. Cytosolic free calcium changes in neutrophils were monitored; (A) C5a was incubated at 37°C in the presence (7 and 20 minutes) and absence of human serum (control); (B) in the absence of C5a (serum only).

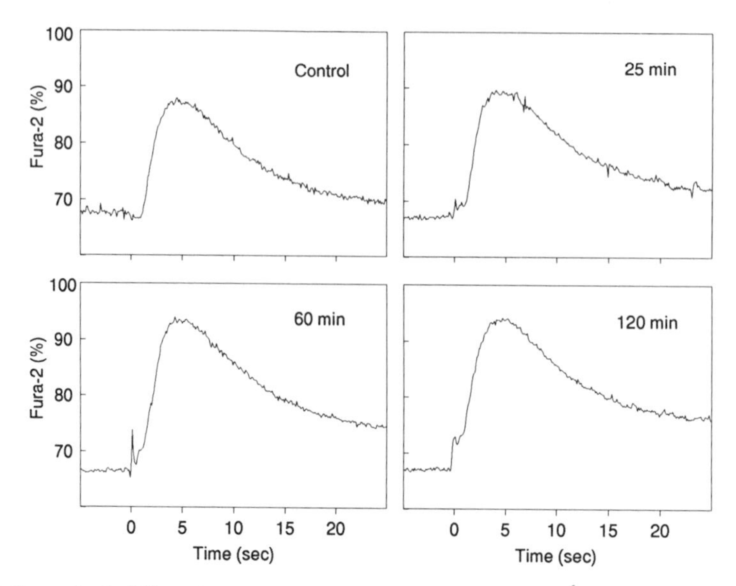

Figure 5. Stability of NAP-2 in human serum. NAP-2 (10^{-6} M) was incubated in the absence (control) and presence of fresh human sera at 37°C for various times. The ability to induce calcium mobilization was tested in fura-2 loaded neutrophils.

recording fura-2 fluorescence changes in human neutrophils (Figure 5). Essentially no decrease in NAP-2 activity was observed during the first two hours. Similar observations were made earlier with NAP-1/IL-8 (8).

POSSIBLE ROLE OF NAP-2 IN PATHOLOGY

We have shown that NAP-2 is a potent activator and attractant for human neutrophils *in vitro* and *in vivo*. NAP-2 closely resembles NAP-1/IL-8 in terms of potency and profile of activity, but has a different origin. NAP1/IL-8 is formed and secreted by mononuclear phagocytes and a variety of tissue cells in response to stimulation with LPS, interleukin-1 or TNF. NAP-2 on the other hand, is a cleavage product of preformed precursors PBP and of CTAP-III, two α-granule proteins that are released from platelets upon activation. Proteases originating from monocytes and possibly from other cellular sources are involved in the processing of these inactive precursors into NAP-2. PBP and CTAP-III released from platelets into circulation are converted by proteolytic cleavage into β-thromboglobulin (β-TG), which is then processed to NAP-2 (Figure 6). Plasma levels of 'β-TG' determined with a radioimmunoassay (which cannot distinguish between the different cleavage products) range from 25 ng/ml in normal individuals to approximatively 200 ng/ml in patients with increased platelet activation. However, one might expect much higher local concentrations of 'β-TG' under conditions where thrombi are formed and blood flow is strongly reduced.

The presence of neutrophils in platelet aggregates has been taken to suggest that activated platelets release chemoattractants. Candidates for this role are both platelet-activating factor and NAP-2. Neutrophils could bring about the degradation of thrombi by the release of proteases such as elastase, and thus lead to the recanalization of the obstructed vessel. On the other hand, it is also possible that neutrophil recruitment and activation could aggravate the course of thrombotic disturbances and atherosclerosis by inducing a local inflammation and tissue damage.

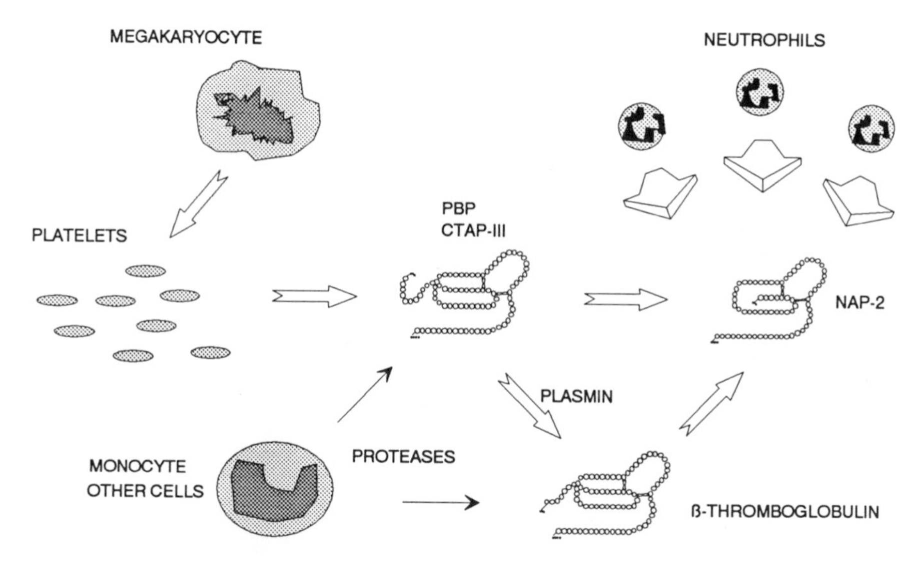

Figure 6. Schematic view of NAP-2 formation.

Cardiopulmonary bypass surgery may result in hemostatic failure and pulmonary endothelial damage which has been attributed to platelet and subsequent neutrophil activation. Indeed, platelet activation is observed when blood is in contact with synthetic surfaces. A 6-fold increase in 'β-TG' and 3-fold rise of elastase activity, paralleled with release of reactive oxygen species, was measured under such conditions (22, 23), indicating an involvement of chemoattractants such as NAP-2. Finally NAP-2 might also induce the lung damage observed in patients with adult respiratory distress syndrome (ARDS). Levels of platelet factor 4 and 'β-TG' are normally undetectable in bronchoalveolar lavage of healthy subjects, but are elevated in bronchoalveolar lavage of patients with ARDS, where the levels of 'β-TG' correlate with the severity of the lung damage (24).

Furthermore NAP-2 might be involved in the formation of pulmonary oedema and cardiogenic shock after atherosclerotic thrombotic and myocardial infarction, where poor outcome and mortality correlates with plasma levels of 'β-TG' (25, 26).

Acknowledgements: This work was supported in part by the Swiss National Science Foundation, Grant No. 31-25700.88 and the Swiss Federal Commission for Rheumatic Diseases.

REFERENCES

1. Walz A., and M. Baggiolini. 1989. A novel cleavage product of β-thromboglobulin formed in cultures of stimulated mononuclear cells activates human neutrophils. *Biochem. Biophys. Res. Commun.* 159: 969-975.
2. Walz A., P. Peveri, H. Aschauer, and M. Baggiolini. 1987. Purification and amino acid sequencing of NAF, a novel neutrophil-activating factor produced by monocytes. *Biochem. Biophys. Res. Commun.* 149: 755-761.
3. Baggiolini M., A. Walz, and S.L. Kunkel. 1989. Neutrophil-activating peptide-1/interleukin-8, a novel cytokine that activates neutrophils. *J. Clin. Invest.* 84: 1045-1049.
4. Anisowicz A., L. Bardwell, and R. Sager. 1987. Constitutive overexpression of a growth-regulated gene in transformed Chinese hamster and human cells. *Proc. Natl. Acad. Sci. USA* 84: 7188-7192.
5. Richmond A., E. Balentien, H.G. Thomas, G. Flaggs, D.E. Barton, J. Spiess, R. Bordoni, U. Francke, and R. Derynck. 1988. Molecular characterization and chromosomal mapping of melanoma growth stimulatory activity, a growth factor structurally related to β-thromboglobulin. *EMBO J.* 7: 2025-2033.

6. Moser B., I. Clark-Lewis, R. Zwahlen, and M. Baggiolini. 1990. Neutrophil-activating properties of the melanoma growth-stimulatory activity. *J. Exp. Med.* 171: 1797-1802.
7. Yoshimura T.K., K. Matsushima, J.J. Oppenheim, and E. Leonard. 1987. Neutrophil chemotactic factor produced by lipopolysaccharide (LPS)-stimulated human blood mononuclear leukocytes: Partial characterization and separation from interleukin 1 (IL-1). *J. Immunol.* 139: 788-793.
8. Peveri P., A. Walz, B. Dewald, M. Baggiolini. 1988. A novel neutrophil-activating factor produced by human mononuclear phagocytes. *J. Exp. Med.* 167: 1547-1560.
9. Holt J.C., M.E. Harris, A.M. Holt, E. Lange, A. Henschen, and S. Niewiarowski. 1986. Characterization of human platelet basic protein, a precursor form of low-affinity platelet factor 4 and β-thromboglobulin. *Biochemistry* 25: 1988-1996.
10. Castor C.W., J.W. Miller, D.A. Walz. 1983. Structural and biological characteristics of connective tissue activating peptide (CTAP-III), a major human platelet-derived growth factor. *Proc. Natl. Acad. Sci. USA* 80: 765-769.
11. Van Damme J., J. Van Beeumen, R. Conings, B. Decock, and A. Billiau. 1989. Purification of granulocyte chemotactic peptide/interleukin-8 reveals N-terminal sequence heterogeneity similar to that of β-thromboglobulin. *Eur. J. Biochem.* 181: 337-344.
12. Deuel T.F., P.S. Keim, M. Farmer, and R.L. Heinrikson. 1977. Amino acid sequence of human platelet factor 4. *Proc. Natl. Acad. Sci. USA* 74: 2256-2258.
13. Walz A., and M. Baggiolini. 1990. Generation of the neutrophil-activating peptide NAP-2 from platelet basic protein or connective tissue activating peptide III through monocyte proteases. *J. Exp. Med.* 171: 449-454.
14. Wenger R.H., A. Wicki, A. Walz, N. Kieffer, and K.J. Clemetson. 1989. Cloning of cDNA coding for connective tissue activating peptide III from a human platelet derived lambda gt11 expression library. *Blood* 73: 1498-1503.
15. Castor C.W., D.A. Walz, P.A. Ragsdale, E.M. Hossler, M.C. Bignall, B.P. Aaron, and K. Mountjoy. 1989. Connective tissue activation. XXXIII. Biologically active cleavage products of CTAP-III from human platelets. *Biochem. Biophys. Res. Commun.* 163: 1071-1078.
16. Walz A., B. Dewald, V. von Tscharner, and M. Baggiolini. 1989. Effects of the neutrophil-activating peptide NAP-2, platelet basic protein, connective tissue activating peptide III and platelet factor 4 on human neutrophils. *J. Exp. Med.* 170: 1745-1750.
17. Schroeder J.M., U. Mrowietz, E. Morita, and E. Christophers. 1987. Purification and partial characterization of a human monocyte-derived, neutrophil-activating peptide that lacks interleukin 1 activity. *J. Immunol.* 139: 3474-3483.
18. Walz A., B. Dewald, and M. Baggiolini. 1990. Formation and biological activity of NAP-2, a neutrophil-activating peptide derived from platelet α-granule precursors, in: *Molecular and cellular biology of cytokines* edited by J.J. Oppenheim, M.C. Powanda, M.J. Kluger and C.A. Dinarello. Alan A. Liss, Inc., New York, in press.
19. Deuel T.F., R.M. Senior, D. Chang, G.L. Griffin, R.L. Heinrikson, and E.T. Kaiser. 1981. Platelet factor 4 is chemotactic for neutrophils and monocytes. *Proc. Natl. Acad. Sci. USA* 78: 4584-4587.
20. Baggiolini M,, B. Dewald. 1984. Exocytosis by neutrophils. *Contemp. Top. Immunobiol.* 14: 221-246.
21. Lindley I., H. Aschauer, J.M. Seiffert, C. Lam, W. Brunowsky, E. Kownatzky, M. Thelen, P. Peveri, B. Dewald, V. von Tscharner, A. Walz, and M. Baggiolini. 1988. Synthesis and expression in Escherichia coli of the gene encoding monocyte-derived neutrophil-activating factor: biological equivalence between natural and recombinant neutrophil-activating factor. *Proc. Natl. Acad. Sci. USA* 85: 9199-9202.
22. Wachtvogel Y.T., U. Kucich, J. Greenplate, P. Gluszko, W. Abrams, G. Weinbaum, R.K. Wenger, B. Rucinski, S. Niewiaroski, L.H. Edmunds Jr., et al. 1987. Human neutrophil degranulation during extracorporeal circulation. *Blood* 69: 324-330.
23. Van Oeveren W., M.D. Kazatchkine, B. Descamps-Latscha, F. Maillet, E. Fischer, A. Carpentier, and C.R. Wildevuur. 1985. Deleterious effects of cardiopulmonary bypass. A prospective study of bubble versus membrane oxygenation. *J. Thorac. Cardiovasc. Surg.* 89: 888-899.
24. Idell S., R. Maunder, A.M. Fein, H.I. Swirtalska, G.P. Tuszynski, J. McLarty, and S. Niewiarowsky. 1989. Platelet-specific α-granule proteins and thrombospondin in bronchoalveolar lavage in the adult respiratory distress syndrome. *Chest* 96: 1125-1132.
25. Oswald G.A., C.C. Smith, A.P. Delamothe, D.J. Betteridge, and J.S. Yudkin. 1988. Raised concentrations of glucose and adrenaline and increased *in vivo* platelet activation after myocardial infarction. *Br. Heart. J.* 59: 663-710.
26. Woo E., C.Y. Huang, V. Chan, Y.W. Chan, Y.L. Yu, and T.K. Chan. 1988. β-thromboglobulin in cerebral infarction. *J. Neurol. Neurosurg. Psychiatry* 51: 557-562.

HUMAN MONOCYTE CHEMOATTRACTANT PROTEIN-1 (MCP-1)

Teizo Yoshimura and Edward J. Leonard

Laboratory of Immunobiology
National Cancer Institute
Frederick, MD 21702, USA

INTRODUCTION

Macrophage accumulation is one of the histological characteristics of delayed-type hypersensitivity (DTH) reactions, chronic inflammation, and certain kinds of tumors. Although the mechanisms of macrophage infiltration into the reaction sites are not fully understood, involvement of macrophage (monocyte) chemotactic factors produced at the reaction sites appear to be important. Those chemotactic factors can be separated into two groups; one is serum protein-derived, the other is cell-derived. A chemotactic factor that may account for massive macrophage infiltration into DTH reaction sites is lymphocyte-derived chemotactic factor (LDCF), which is produced in various species by antigen- or mitogen-stimulated spleen cells or peripheral blood mononuclear cells (PBMC) (1). Tumor cell-derived chemoattractants with similar physico-chemical characteristics have also been described, but have not been purified (2,3). On the other hand, several cytokines that were identified and purified based on their biological activities other than chemotactic activity have been reported to be chemotactic for leukocytes (4-8). Interleukin 1 was among those (4). In 1987, we succeeded in separating neutrophil chemotactic activity from IL-1 activity in the culture supernatant of LPS-stimulated human PBMC (9), and then purified the protein (10). It was initially termed monocyte-derived neutrophil chemotactic factor (MDNCF), and is now called neutrophil attractant/activation protein-1 (NAP-1). In the same supernatant we also found monocyte chemotactic activity (MCA). However, we were unable to purify the protein because of the limited supply of the supernatant. Fortunately, human malignant glioma cell lines were found to produce large amounts of MCA, with physico-chemical characteristics very similar to those of mitogen-stimulated PBMC-derived MCA (11). This discovery made it possible to purify, sequence, and clone the protein, which we call monocyte chemoattractant protein-1 (MCP-1).

PURIFICATION OF MCP-1

MCP-1 was purified from serum-free U-105MG culture fluid in 3 steps: dye-ligand chromatography on Orange-A Agarose, CM-HPLC chromatography, and RP-HPLC (12). MCP-1 is a relatively basic protein, as shown by its binding to a cation exchange column at pH 6.5 and elution from a chromatofocusing column

Chemotactic Cytokines, Edited by J. Westwick *et al.*
Plenum Press, New York, 1991

between pH 7.5 and 9.2 (11,12). The cation exchange column separated MCP-1 into two well-separated peaks.

The two proteins behaved as distinct entities on SDS-PAGE, migrating as 15 and 13 kD proteins (Figure 1, MCP-1α and -β). However, elution times of MCP-1α and -β from the RP-HPLC column were identical; and since amino acid compositions of the two proteins were also indistinguishable, we believe that the two MCP-1's represent a single gene product, differences being due to post-translational modifications as yet unidentified.

In relation to the possibility of glycosylation, both MCP-1α and -β were secreted by PDGF-stimulated fibroblasts, even in the presence of tunicamycin, an inhibitor of N-linked glycosylation (13).

Additional support for gene product identity of MCP-1α and -β is the fact that both forms were precipitated by a rabbit antibody induced by pure MCP-1β (13). The molecular mass of MCP-1 is 8700 daltons, based on amino acid sequence and cloning data (see below). Thus the SDS-PAGE result noted above is anomalous. A similar discrepancy was described for MGSA, which comprises 73 amino acids but migrates as a 13 kD species (14).

MCP-1α and -β have also been isolated from culture fluid of PHA-stimulated human blood mononuclear cells (15). In contrast to glioma cell culture supernatant, however, MCP-1 accounts for only half of the total MCA in the PHA-stimulated PBMC culture supernatant, suggesting that mitogen-stimulated PBMC produce other chemoattractants as well. After immunoprecipitation of malignant fibrous histiocytoma cell line culture supernatants, we recently found another band which migrated in SDS-PAGE to a point just above MCP-1α (16). Heterogeneity of MCP-1 has been also reported by others.

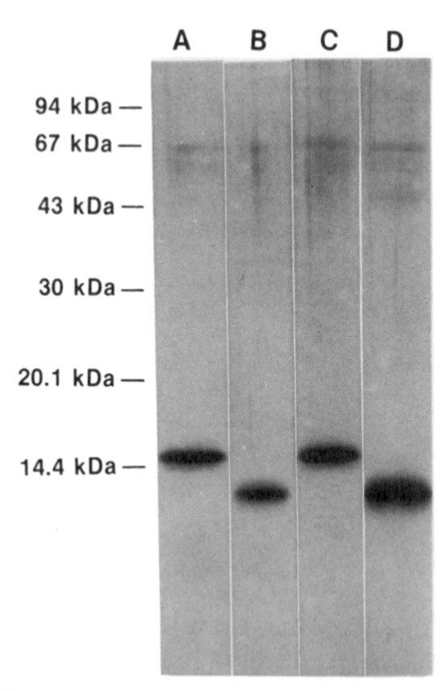

Figure 1. SDS-PAGE of MCP-1 purified from conditioned media of glioma cells or PHA-stimulated human blood mononuclear cells. Approximately 25 ng of protein was applied to a 15% polyacrylamide gel under reducing conditions. Lanes A and B: glioma-derived MCP-1α and -β. Lanes C and D: mononuclear cell-derived MCP-1α and -β.

SEQUENCING AND cDNA CLONING

MCP-1β comprises a single 76 residue protein chain. Since the N-terminus is blocked (pyroglutamic acid), sequencing required analysis of the products of proteolytic cleavage, which was achieved by a combination of Edman degradation and mass spectrometry (17). The sequence was confirmed by full-length cDNA cloning of MCP-1 from a glioma cell line library (18). The cDNA open reading frame codes for a 99 residue protein. The last 76 residues correspond to MCP-1. Hydrophobicity of the first 23 residues is typical of a signal peptide, which is consistent with the fact that MCP-1 is a secreted protein. There is a single consensus sequence (Asn-Phe-Thr) for N-linked glycosylation targeting amino acid 38. The A+T content of the 3'-untranslated region (66%) is not nearly as high as that found in some mRNAs that are transiently expressed in response to a cell activation stimulus. Unlike a number of genes encoding proteins related to the inflammatory response (19), there is no 8-nucleotide TTATTTAT sequence in the 3'-untranslated region. Analysis of restriction endonuclease digests of human genomic DNA shows that there is a single MCP-1 gene. Under conditions of high stringency, MCP-1 cDNA hybridized with genomic DNA restriction fragments from chimpanzee and baboon, but not rat or mouse (18).

There are 4 half-cystines in MCP-1, at positions 11, 12, 36 and 52. Analysis of MCP-1α shows that they exist as disulfides, since they can be carboxymethylated

```
                                 -50       -40       -30       -20       -10        -1
                                   CTAACCCAGAAACATCCAATTCTCAAACTGAAGCTCGCACTCTCGCCTCCAGC

                10          20          30          40          50          60
ATG AAA GTC TCT GCC GCC CTT CTG TGC CTG CTG CTC ATA GCA GCC ACC TTC ATT CCC CAA
Met Lys Val Ser Ala Ala Leu Leu Cys Leu Leu Leu Ile Ala Ala Thr Phe Ile Pro Gln

                70          80          90         100         110         120
GGG CTC GCT CAG CCA GAT GCA ATC AAT GCC CCA GTC ACC TGC TGT TAT AAC TTC ACC AAT
Gly Leu Ala Gln Pro Asp Ala Ile Asn Ala Pro Val Thr Cys Cys Tyr Asn Phe Thr Asn
           △                                                    - - - - - - - -

               130         140         150         160         170         180
AGG AAG ATC TCA GTG CAG AGG CTC GCG AGC TAT AGA AGA ATC ACC AGC AGC AAG TGT CCC
Arg Lys Ile Ser Val Gln Arg Leu Ala Ser Tyr Arg Arg Ile Thr Ser Ser Lys Cys Pro

               190         200         210         220         230         240
AAA GAA GCT GTG ATC TTC AAG ACC ATT GTG GCC AAG GAG ATC TGT GCT GAC CCC AAG CAG
Lys Glu Ala Val Ile Phe Lys Thr Ile Val Ala Lys Glu Ile Cys Ala Asp Pro Lys Gln

               250         260         270         280         290         300
AAG TGG GTT CAG GAT TCC ATG GAC CAC CTG GAC AAG CAA ACC CAA ACT CCG AAG ACT TGA
Lys Trp Val Gln Asp Ser Met Asp His Leu Asp Lys Gln Thr Gln Thr Pro Lys Thr
                        ___________________________________

        310       320       330       340       350       360       370       38
ACACTCACTCCACAACCCAAGAATCTGCAGCTAACTTATTTTCCCCTAGCTTTCCCCAGACACCCTGTTTTATTTTATT

0        390       400       410       420       430       440       450       4
ATAATGAATTTTGTTTGTTGATGTGAAACATTATGCCTTAAGTAATGTTAATTCTTATTTAAGTTATTGATGTTTTAAG

60        470       480       490       500       510       520       530
TTTATCTTTCATGGTACTAGTGTTTTTTAGATACAGAGACTTGGGGAAATTGCTTTTCCTCTTGAACCACAGTTCTACC

540       550       560       570       580       590       600       610
CCTGGGATGTTTTGAGGGTCTTTGCAAGAATCATTAATACAAAGAATTTTTTTTAACATTCCAATGCATTGCTAAAATA

 620       630       640       650       660       670       680
TTATTGTGGAAATGAATATTTTGTAACTATTACACCAAATAAATATATTTTTGTACAAAAAAAAAAAAAA
                                     ......
```

Figure 2. Nucleotide sequence of human MCP-1. Triangle: N-terminus of mature MCP-1. Dashed line: potential N-linked glycosylation site. Solid line: Oligonucleotide probe sequence. Dotted line: polyadenylation signal.

only after reduction. The 4 half-cystines are the basis for assigning MCP-1 to a family of proteins with amino acid sequence similarity ranging from 28% to 55% and half-cystines in identical locations (18). A related family of proteins, of which NAP-1 is a member, differs from the MCP-1 family in that the first 2 of the 4 half-cystines are separated by a single amino acid residue (C-X-C). Structural studies showed that β-thromboglobulin, a member of the C-X-C family, has two disulfide bridges, formed by the first and third and by the second and fourth C's in the sequence (20). These bridges create two loops; the disulfides form the bases of the loops, and the bases are close together because of the proximity of the first two C's. Reduction and alkylation of NAP-1 caused loss of biological activity, which suggests that the oxidized state is the functionally active structure (21). Since neither loop of NAP-1 alone had chemotactic activity (21), it is likely that the paired disulfide loops are required for biological activity, and the same requirements may apply to MCP-1. These leukocyte-specific attractants, NAP-1 and MCP-1, may have evolved from a primitive molecule, the paired disulfide loops of which determined the configuration that was required for interaction with the cell receptor. Within constraints imposed by the loops, unique leukocyte specificity of the two molecules developed, via differences in primary sequence (only 24% sequence similarity between NAP-1 and MCP-1).

BIOLOGICAL ACTIVITY OF MCP-1 *IN VITRO*

MCP-1 is a chemoattractant for human monocytes. Approximately 30% of input monocytes respond at the optimal agonist concentration of 10^{-9}M (12,15). This value is comparable to that found for other chemoattractants (22). The migratory response to MCP-1 is primarily chemotactic rather than chemokinetic. Human neutrophils do not migrate to MCP-1 concentrations as high as 3×10^{-8}M (12,15). Basophils respond to MCP-1; there is no detectable activity for eosinophils. In contrast to the respiratory burst induced by formyl peptide attractants (23), MCP-1 causes minimal or no release of monocyte superoxide (Leonard *et al*, unpublished).

It is well known that macrophages become competent to perform certain functions only after activation by specific signals (24). It was shown that activation of peritoneal macrophages to kill syngeneic tumor cells without the intervention of antibody required two signals (MAF or IFN-gamma, and LPS), which could be applied either together or sequentially, provided LPS was second (25). Activation of macrophages for other functions may require different signals. For example, competence of mouse resident peritoneal macrophages to respond to the chemoattractant C5a can be induced by macrophage stimulating protein (MSP) (26), a large macromolecule unrelated to known cytokines (Skeel *et al.*, unpublished). It was recently shown that when human tumor cells were added to monolayers of human monocytes, thymidine incorporation after 68 h in culture was inhibited if MCP-1 was added at the beginning of the culture period. Not all tested tumor cell lines were affected (27). This report suggests that it will be of interest to determine if MCP-1, alone or in combination with other cytokines, not only attracts monocytes to the site of its release but also causes cellular activation for specific functions related to host defense.

IDENTIFICATION OF MCP-1 RECEPTOR

The binding of MCP-1 to human monocytes was studied with iodinated MCP-1 (^{125}I-MCP-1) (28). Iodination was performed with IODO-BEADS without significant loss of chemotactic activity. The binding of ^{125}I-MCP-1 to PBMC

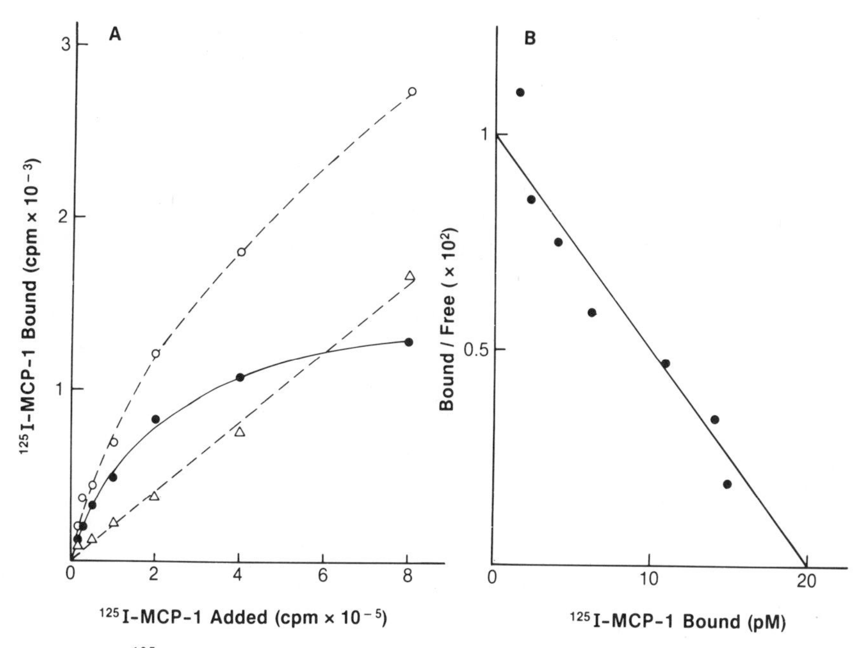

Figure 3. ^{125}I-MCP-1 binding curve and Scatchard plot analysis. A. Binding curve. Open circle: Total binding. Triangle: nonspecific binding. Closed circle: specific binding. B. Scatchard plot analysis.

occurred within 10 minutes at 0°C. The binding was dose-dependently inhibited by unlabelled MCP-1 but not by fMLP or NAP-1, suggesting that there are MCP-1 specific receptors. So far monocyte is the only cell population that expresses MCP-1 receptors on the surface. Scatchard plot analysis indicated that monocytes had a minimum of 1700 ± 600 binding sites per cell with a Kd of $1.9 \pm 0.2 \times 10^{-9}$ M. In order to investigate whether or not there are monocyte subpopulations in terms of MCP-1 binding, MCP-1 was biotinylated. Biotinylation resulted in a loss of potency of chemotactic activity, but efficacy was retained. FACS analysis was done with this biotinylated MCP-1. Although significant binding (inhibited by excess unlabelled MCP-1) was observed, the histogram was too close to the control to determine if there is a subpopulation of monocytes that does not bind MCP-1. Binding of biotinylated MCP-1 to lymphocytes was not seen.

CELLULAR SOURCES OF MCP-1 AND POSSIBLE ROLES *IN VIVO*

As mentioned above, MCP-1 belongs to a family of proteins with molecular masses of about 10 kD and four half-cystines at almost identical positions. Among them, mouse JE had the highest sequence similarity to human MCP-1 (55% amino acid and 68% nucleotide sequence similarity) (18). JE had been defined 6 years earlier as a cDNA cloned from mouse fibroblasts after stimulation by PDGF (29). Although the gene product was not identified and its function was therefore unknown, the cell sources of JE and the stimuli for its expression were studied extensively. Recently, human JE was cloned from human fibroblasts and it was found that human JE was identical to human MCP-1 (30). Discovery of the identity of MCP-1 and JE gene products makes it possible to relate MCP-1 to the considerable body of previous work on the generation of JE mRNA in response to various activation stimuli (31-36). Table 1 summarizes our present knowledge, based on either increases in cell JE mRNA or identification of MCP-1 protein. Both glioma cell lines (11,12) and baboon aortic smooth muscle cells (44) secrete MCP-1 without addition of known stimuli. Some glioma lines secrete PDGF - which acts as an autocrine stimulus for glioma replication (47) and could stimulate JE expression. As for production of MCP-1 by aortic smooth muscle cells, the stimulus might be provided by FCS in the culture medium. Although the concentration of PDGF in FCS is low compared to human serum, mitogenic activities of these two sera are comparable (48), suggesting the presence of other mitogenic factors in FCS that might also stimulate JE expression. This idea is supported by our finding that human fibroblast cell lines in media containing FCS secrete MCP-1 without additional stimuli. We have also shown that addition of PDGF to confluent cultures of these cells in the absence of FCS causes secretion of MCP-1α and -β (13).

Table 1. Cellular Expression of MCP-1 or JE mRNA.

Cell	Species	Stimulus
Lymphocyte	human	PHA (15,18,31), IL-1 (18), LPS (18)
Macrophage	mouse	LPS (32), Listeria (32), IFN-gamma (36)
Monocyte	human	LPS ? (30,37)
Fibroblast	mouse	PDGF (29,33), IL-1 (32), VSV (34), Poly(I):poly(C) (34)
	human	IL-1, TNF (38,39), PDGF (13) Poly(I):poly(C), MV, RV (40)
Endothelium	rat	EGF (35)
	human	IL-1 (39,42), TNF (39,41,42), LPS (39), TGF-β (43)
Smooth muscle	baboon	(-) (44)
Tumor cells	human	
Malignant glioma		(-) (11,12,18,45)
THP-1		LPS + silica + hydroxyurea (27)
HL60		LPS (46)
Melanoma		(-) (40,45)
Fibrosarcoma		(-) (45)
Osteosarcoma		(-) (40,45)
MFH		(-) (16)

Abbreviations: VSV, vesicular stomatitis virus; poly(I):poly(C), double-stranded RNA; MV, measles virus; RV, rubella virus; EGF, epidermal growth factor; TNF, tumor necrosis factor. MFH, malignant fibrous histiocytoma.

MCP-1 can be produced by leukocytes of both lymphocyte and monocyte lineage. Addition of PHA to human blood mononuclear cells (monocytes and lymphocytes) stimulated MCP-1 production, as shown by Northern blotting (18,31) and by isolation of MCP-1 (15). In this mixed population, the cell expressing mRNA for MCP-1 was the T-lymphocyte. Lymphocytes depleted of monocytes did not express MCP-1 mRNA in response to PHA (31), which reflects a requirement for a small number of monocytes in lymphocyte responses to mitogen (49). Addition of LPS to human blood mononuclear cells stimulated expression of MCP-1 mRNA (18). Since LPS activates monocytes, and probably does not have a direct effect on lymphocytes, it will be important to determine whether the monocyte secretes MCP-1 in response to LPS or secretes a stimulus for lymphocyte MCP-1 production. It would be reasonable for differentiated tissue macrophages, which can be regarded as one of the first lines of host defense, to secrete MCP-1 in response to noxious stimuli. In this context, addition of LPS to mouse peritoneal macrophages caused expression of JE mRNA (32). MCP-1 was isolated from the culture supernatant of LPS-stimulated human monocytes. Rollins *et al* have reported that human blood monocytes, purified from mononuclear cell preparations by E-rosette removal of lymphocytes, secrete MCP-1 without additional stimuli (41). This is a puzzling finding, which might be related to the E-rosette procedure.

Among other cellular sources, MCP-1 is produced by endothelial cells in response to various stimuli (35,39,41-43). Valente *et al.* purified a monocyte chemotactic factor from the culture supernatant of baboon aortic smooth muscle cells (SMC-CF) (44). Although they did not show amino acid sequence of SMC-CF, the amino acid composition data was almost identical to that of human MCP-1. Hybridization of the MCP-1 cDNA probe with baboon genomic DNA is added evidence for the relation between MCP-1 and SMC-CF (18). Penetration of vascular endothelium by blood monocytes occurs in experimental atherosclerosis (50). Since local production of MCP-1 could mediate accumulation of monocytes in the lesions, it will be interesting to determine by in situ hybridization if MCP-1 mRNA is detectable in the lesion in the early stage of atherosclerosis.

There is an extensive literature on infiltration of tumors by macrophages and destruction of tumor cells by activated macrophages (51). Accumulation of macrophages at tumor sites could be mediated by attractants generated directly by the tumor or indirectly by a cellular immune reaction to tumor antigen.

Results summarized here show that MCP-1 can be elaborated by both mechanisms (assuming that the response to PHA is a prototype of an antigen-driven reaction). Although several glioma cell lines produced MCP-1, Northern blot analysis of a number of other human tumor cell lines failed to reveal MCP-1 message (8). It is possible that secreted MCP-1 might have been found in culture fluids of some of these lines. Chemotactic activity for mouse macrophages was reported for culture fluids of 5 different chemically induced mouse sarcoma lines (2). Although this activity was only partially purified, its molecular mass of about 15 kD and its inability to attract mouse granulocytes suggest similarity to MCP-1. Human tumor cell-derived monocyte chemotactic factor has also been reported (3). Although TDCF has not been purified, Bottazzi *et al.* have concluded that TDCF is identical to MCP-1, based on the fact that various kinds of tumor cells produce MCP-1 and MCP-1 mRNA is expressed by tumor cells used for TDCF studies (52). Accumulation of macrophages at tumor sites, whether mediated by host or tumor derived attractants, is a necessary but not sufficient condition for tumor destruction. Both cytopathic and trophic roles have been attributed to infiltrating tumor macrophages.

As noted above, macrophage-mediated tumor cell destruction requires macrophage activation, and the possible role of MCP-1 as an activator merits further study.

CONCLUSION

The purification, complete amino acid sequencing, and cDNA cloning have led us to a new stage of investigation in this field. As shown here, we are aware that many kinds of cells can produce MCP-1 with or without a stimulus *in vitro*. However, the accumulation of macrophages is delicately regulated *in vivo*. Therefore, our future studies will be to determine if MCP-1 is really responsible for *in vivo* macrophage infiltration, and if so, to find what cells are responsible for the production of MCP-1. Development of specific antibody against MCP-1 and establishment of an animal model will be helpful for future experiments.

REFERENCES

1. Altman, L. C. 1978. Chemotactic lymphokines. In *Leukocyte Chemotaxis*. Gallin, J.I. and Quie, P.G., eds. Raven Press, N.Y. p 267-279.
2. Meltzer, M. S., M. M. Stevenson, and E. J. Leonard. 1977. Characterization of macrophage chemotaxins in tumor cell cultures and comparison with lymphocyte-derived chemotactic factor. *Cancer Res*. 37: 721-725.
3. Bottazzi, B., N. Polentarutti, R. Acero, A. Balsari, D. Boraschi, P. Ghezzi, M. Salmona, and A. Mantovani. 1983. Regulation of macrophage content of neoplasms by chemoattractant. *Science* 220: 210-212.
4. Luger, T. A., J. A. Charon, M. Colot, M. Micksche, and J. J. Oppenheim. 1983. Chemotactic properties of partially purified human epidermal cell-derived thymocyte-activating factor (ETAF) for polymorphonuclear and mononuclear cells. *J. Immunol.* 131: 816-820.
5. Ming, W. J., L. Bersani, and A. Mantovani. 1987. Tumor necrosis factor is chemotactic for monocytes and polymorphonuclear leukocytes. *J. Immunol.* 138: 1469-1474.
6. Wang, J. M., S. Colella, P. Allavena, and A. Mantovani. 1987. Chemotactic activity of human recombinant granulocyte-macrophage colony-stimulating factor. *Immunology* 60: 439-444.
7. Wang, J. M., J. D. Griffin, A. Rambaldi, Z. G. Chen, and A. Mantovani. 1988. Induction of monocyte migration by recombinant macrophage colony-stimulating factor. *J. Immunol.* 141: 575-579.
8. Wahl, S. M., D. A. Hunt, L. M. Wakefield, N. McCartney-Francois, L. M. Wahl, A. B. Roberts, and M. B. Sporn. 1987. Transforming growth factor type beta induces monocyte chemotaxis and growth factor production. *Proc. Natl. Acad. Sci. USA*. 84: 5788-5792.
9. Yoshimura, T., K. Matsushima, J. J. Oppenheim, and E. J. Leonard. 1987. Neutrophil chemotactic factor produced by lipopolysaccharide (LPS)-stimulated human blood mononuclear leukocytes: partial characterization and separation from interleukin 1 (IL 1). *J. Immunol.* 139: 788-793.
10. Yoshimura, T., K. Matsushima, S. Tanaka, E. A. Robinson, E. Appella, J. J. Oppenheim, and E. J. Leonard. 1987. Purification of a human monocyte-derived neutrophil chemotactic factor that has peptide sequence similarity to other host defense cytokines. *Proc. Natl. Acad. Sci. USA*. 84: 9233-9237.
11. Kuratsu, J., E. J. Leonard, and T. Yoshimura. 1989. Production and characterization of human glioma-derived monocyte chemotactic factor. *J. Natl. Cancer Inst.* 81: 347-351.
12. Yoshimura, T., E. A. Robinson, S. Tanaka, E. Appella, J. Kuratsu, and E. J. Leonard. 1989. Purification and amino acid analysis of two human glioma-derived monocyte chemoattractants. *J. Exp. Med.* 169: 1449-1459.
13. Yoshimura, T., and E. J. Leonard. 1990. Secretion by human fibroblasts of monocyte chemoattractant protein-1 (MCP-1), the product of gene JE. *J. Immunol.* 144: 2377-2383.
14. Richmond, A., E. Balentien, H. G. Thomas, G. Flaggs, D. E. Barton, J. Spiess, R. Bordoni, U. Francke, and R. Derynck. 1988. Molecular characterization and chromosomal mapping of melanoma growth stimulating activity, a growth factor structurally related to β-thromboglobulin. *EMBO J.* 7: 2025-2033.
15. Yoshimura, T., E. A. Robinson, S. Tanaka, E. Appella, and E. J. Leonard. 1989. Purification and amino acid analysis of two human monocyte chemoattractants produced by phytohemagglutinin-stimulated human blood mononuclear leukocytes. *J. Immunol.* 142: 1956-1962.
16. Takeya, M., T. Yoshimura, E. J. Leonard, T. Kato, H. Okabe, and K. Takahashi. Production of monocyte chemotactic protein-1 (MCP-1) by malignant fibrous histiocytoma. *Submitted.*
17. Robinson, E. A., T. Yoshimura, E. J. Leonard, S. Tanaka, P. R. Griffin, J. Shabanowitz, D. F. Hunt, and E. Appella. 1989. Complete amino acid sequence of a human monocyte chemoattractant, a putative mediator of cellular immune reactions. *Proc. Natl. Acad. Sci. USA*. 86: 1850-1854.
18. Yoshimura, T., N. Yuhki, S. K. Moore, E. Appella, M. I. Lerman, and E. J. Leonard: Human monocyte chemoattractant protein-1 (MCP-1). 1989. Full-length cDNA cloning, expression in mitogen-stimulated blood mononuclear leukocytes, and sequence similarity to mouse competence gene JE. *FEBS Letters* 244: 487-493.
19. Caput, D., B. Beutler, K. Hartog, R. Thayer, S. Brown-Shimer, and A. Cerami. 1986. Identification of a common nucleotide sequence in the 3'-untranslated region of mRNA molecules specifying inflammatory madiators. *Proc. Natl. Acad. Sci. USA*. 83: 1670-1674.
20. Begg, G. S., D. S. Pepper, C. N. Chesterman, and F. J. Morgan. 1978. Complete covalent structure of human β-thromboglobulin. *Biochemistry* 17: 1739-1744.

21. Tanaka, S., E. A. Robinson, T. Yoshimura, K. Matsushima, E. J. Leonard, and E. Appella. 1988. Synthesis and biological characterization of monocyte-derived neutrophil chemotactic factor. *FEBS Letters* 236: 467-470.
22. Rot, A., L. E. Henderson, T. D. Copeland, and E. J. Leonard. 1987. A Series of six ligands for the human formyl peptide receptor: tetra peptides with high chemotactic potency and efficacy. *Proc. Natl. Acad. Sci. USA.* 84: 7967-7971.
23. Leonard, E.J., A. Shenai, and A. Skeel. 1987. Dynamics of chemotactic peptide-induced superoxide generation by human monocytes. *Inflammation.* 11: 229-240.
24. Adams, D.O., and T. A. Hamilton. 1988. Activation of macrophages for tumor cell kill: Effector mechanisms and regulation. In Heppner, G.H. and Fulton, A.M. *Macrophages and Cancer,* CRC, Press Inc., Boca Raton, FL., pp 27-38.
25. Ruco, L.P., and M. S. Meltzer. 1978. Macrophage activation for tumor cytotoxicity: Development of macrophage cytotoxic activity requires completion of short-lived intermediary reactions. *J. Immunol.* 121: 2035-2042.
26. Leonard, E.J. and A. Skeel. 1978. Isolation of macrophage stimulation protein (MSP) from human serum. *Exp. Cell Res.* 114: 117-126.
27. Matsushima, K., C. G. Larsen, G. C. DuBois, and J. J. Oppenheim. 1989. Purification and characterization of a novel monocyte chemotactic and activating factor produced by a human myelomonocytic cell line. *J. Exp. Med.* 169: 1485-1490.
28. Yoshimura, T., and E. J. Leonard. 1990. Identification of high affinity receptors for human monocyte chemoattractant protein-1 on human monocytes. *J. Immunol.* in press.
29. Cochran, B. J., A. C. Reffel, and C. D. Stiles. 1983. Molecular cloning of gene sequences regulated by platelet-derived growth factor. *Cell* 33: 939-947.
30. Rollins, B. J., P. Stier, T. Ernst, and G. G. Wong. 1989. The human homologue of the JE gene encodes a monocyte secretory protein. *Mol. Cell Biol.* 9: 4687-4695.
31. Kaczmarek, B., B. Calabretta, and R. Baserga. 1985. Expression of cell-cycle-dependent genes in phytohemagglutinin-stimulated human lymphocytes. *Proc. Natl. Acad. Sci. USA.* 82: 5375-5379.
32. Introna, M., R. C. Bast, Jr., C. S. Tannenbaum, T. A. Hamilton, and D. O. Adams. 1987. The effect of LPS on expression of the early "competence" genes JE and KC in murine peritoneal macrophages. *J. Immunol.* 138: 3891-3896.
33. Rollins, B. J., E. D. Morrison, and C. D. Stiles. 1988. Cloning and expression of JE, a gene inducible by platelet-derived growth factor and whose product has cytokine properties. *Proc. Natl. Acad. Sci. USA.* 85: 3738-3742.
34. Zullo, J.N., B. H. Cochran, A. S. Huang, and C. D. Stiles. 1985. Platelet-derived growth factor and double-stranded ribonucleic acids stimulate expression of the same genes in 3T3 cells. *Cell* 43: 793-800.
35. Takehara, K., E. C. LeRoy, and G. R. Grotendorst. 1987. TGF-β inhibition of endothelial cell proliferation: alteration of EGF binding and EGF-induced growth-regulatory (competence) gene expression. *Cell* 49: 415-422.
36. Prpic, V., S-F. Yu, F. Figueiredo, P. W. Hollenbach, G. Gawdi, B. Herman, R. J. Uhing, and D. O. Adams. 1989. Role of Na^+/H^+ exchange by interferon-gamma in enhanced expression of JE and I-A_β genes. *Science* 244: 469-471.
37. Decock, B., R. Conings, J-P. Lenaeerts, A. Billiau, and J. Van Damme. 1990. Identification of the monocyte chemotactic protein from human osteosarcoma cells and monocytes: Detection of a novel N-terminally processed form. *Biochem. Biophys. Res. Commun.* 167: 904-909.
38. Larsen, C. G., C. O. C. Zachariae, J. J. Oppenheim, and K. Matsushima. 1989. Production of monocyte chemotactic and activating factor (MCAF) by human dermal fibroblasts in response to interleukin 1 or tumor necrosis factor. *Biochem. Biophys. Res. Commun.* 160: 1403-1408.
39. Strieter, R. M., R. Wiggins, S. H. Phan, B. L. Wharram, H. J. Showell, D. G. Remick, S. W. Chensue, and S. L. Kunkel. 1989. Monocyte chemotactic protein gene expression by cytokine-treated human fibroblasts and endothelial cells. *Biochem. Biophys. Res. Commun.* 162: 694-700.
40. Van Damme, J., B. Decock, J-P. Lenaerts, R. Conings, R. Bertini, A. Mantovani, and A. Billiau. 1990. Identification by sequence analysis of chemotactic factors for monocytes produced by normal and transformed cells stimulated with virus, double-stranded RNA or cytokine. *Eur. J. Immunol.* 19: 2367-2373.
41. Dixit, V., S. Green, V. Sarma, L. B. Holzman, F. W. Wolf, K. O'Rourke, P. A. Ward, E. V. Prochownik, and R. M. Marks. 1990. Tumor necrosis factor-α induction of novel gene products in human endothelial cells including a macrophage-specific chemotaxin. *J. Biol. Chem.* 265: 2973-2978.

42. Sica, A., J. M. Wang, F. Colotta, E. Dejana, A. Mantovani, J. J. Oppenheim, C. G. Larsen, C. O. C. Zachariae, and K. Matsushima. 1990. Monocyte chemotactic and activating factor gene expression induced in endothelial cells by IL-1 and tumor necrosis factor. *J. Immunol.* **144:** 3034-3038.
43. Rollins, B. J., T. Yoshimura, E. J. Leonard, and J. S. Pober. 1990. Cytokine-activated human endothelial cells synthesize and secrete a monocyte chemoattractant, MCP-/JE. *Am. J. Pathol.* in press.
44. Valente, A. J., D. T. Graves, C. E. Vialle-Valentin, R. Delgado, and C. J. Schwartz. 1988. Purification of a monocyte chemotactic factor secreted by nonhuman primate vascular cells in culture. *Biochemistry* 27: 4162-4168.
45. Graves, D. T., Y. L. Jiang, M. J. Williamson, and A. J. Valente. 1989. Identification of monocyte chemotactic activity produced by malignant cells. *Science* 245: 1490-1493.
46. Furutani, Y., H. Nomura, M. Notake, Y. Oyamada, T. Fukui, M. Yamada, C. G. Larsen, J. J. Oppenheim, and K. Matsushima. 1989. Cloning and sequencing of the cDNA for human monocyte chemotactic and activating factor (MCAF). *Biochem. Biophys. Res. Commun.* 159: 249-255.
47. Nister, M., A. Hammacher, K. Mellstrom, A. Siegbahn, L. Ronnstrand, B. Westermark, and C-H. Heldin. 1988. A glioma-derived PDGF A chain homodimer has different functional activities from a PDGF AB heterodimer purified from human platelets. *Cell* 52: 791-799.
48. Bowen-Pope, D.F., C. E. Hart, and R. A. Seifert. 1989. Sera and conditioned media contain different isomer of platelet-derived growth factor (PDGF) which bind to different classes of PDGF receptor. *J. Biol. Chem.* 264: 2502-2508.
49. Rosenstreich, D.L., J. J. Farrar, and S. Dougherty. 1976. Absolute macrophage dependency of T lymphocyte activation by mitogens. *J. Immunol.* 116: 131-139.
50. Gerrity, R.G., H. K. Naito, M. Richardson, and C. J. Schwartz. 1979. Dietary induced atherogenesis in swine. *Am. J. Pathol.* 95: 775-792.
51. Fulton, A.M. 1988. Tumor associated macrophages. In *Macrophages and Cancer*, Heppner, G.H. and Fulton, A.M., eds. CRC Press, Boca Raton, FL. pp 97-111.
52. Bottazzi, B., F. Colotta, A. Sica, N. Nobili, and A. Mantovani. 1990. A chemoattractant expressed in human sarcoma cells (tumor-derived chemotactic factor, TDCF) is identical to monocyte chemoattractant protein-1/monocyte chemotactic and activating factor (MCP/ MCAF). *Int. J. Cancer* 45: 795-797.

BIOLOGICAL ASPECTS OF MONOCYTE CHEMOATTRACTANT PROTEIN-1 (MCP-1)

Edward J. Leonard, Alison Skeel, and Teizo Yoshimura

Immunopathology Section
Laboratory of Immunobiology
National Cancer Institute
Frederick, Maryland, USA

INTRODUCTION

In 1973, Altman *et al.* published a paper entitled "A human mononuclear leukocyte chemotactic factor: characterization, specificity and kinetics of production by homologous leukocytes" (1). They reported that stimulation by tuberculin (PPD) of blood mononuclear leukocytes from individuals with a positive skin test to tuberculin caused production of a chemotactic factor for human monocytes. Since addition of PPD to leucocytes from PPD-negative subjects did not elicit the factor, the authors suggested that it was an *in vitro* correlate of delayed hypersensitivity and could account for recruitment of macrophages in cellular immune reactions. The factor was a heat stable macromolecule with a molecular mass, estimated by gel filtration, of about 12,500 daltons. Altman later referred to this molecule as leukocyte-derived chemotactic factor, or LDCF (2). In 1989, we reported the purification to homogeneity of the predominant chemotactic activity for monocytes in culture fluids of PHA-stimulated human mononuclear leukocytes (3). We suggested that this 8700 dalton protein, called MCP-1, is the attractant studied by Altman *et al.* fifteen years earlier. The present communication addresses the possible role of MCP-1 in cellular immunity. Other biological aspects of MCP-1 were recently reviewed (4).

EVALUATING MCP-1 IN DELAYED HYPERSENSITIVITY (DH)

Immune reactions mediated by cells, without necessary participation by antibody, occur in many different conditions that range from tuberculosis to contact allergy (Table 1).

Specific cellular immunity can demonstrated by delayed cutaneous hypersensitivity (DCH) to intradermal injection of antigen. The characteristics of the DCH reaction are outlined in Table 2.

The reaction is delayed in time, developing over a period of 24-48 hours. Erythema reflects local vasodilatation. Swelling is due to deposition of fibrin, generated in response to tissue macrophage procoagulant production, and to cellular infiltration. In contrast to the complement-mediated acute inflammation associated with antigen-antibody reactions, neutrophils are absent unless the reaction is so

Chemotactic Cytokines, Edited by J. Westwick *et al.*
Plenum Press, New York, 1991

Table 1. Stimuli of Cellular Immune Reactions

bacteria and fungi:	M. tuberculosis, Histoplasma, Candida
viruses:	vaccinia, mumps, herpes
insect bites:	mosquito, fleas, ticks
contactants:	DNCB, picryl chloride, urushiol
allografts	
tumors:	both primary and syngeneic transplants

Table 2. Delayed Cutaneous Hypersensitivity [DCH]

24-48 hrs after challenge: red lump
Cellular infiltrate:
mononuclear cells [lymphocytes and monocytes]
basophils
no neutrophils [unless necrosis]
Fibrin deposition [macrophage procoagulant]

severe that necrosis occurs. The relative contribution of monocytes and lymphocytes to the predominantly mononuclear cell infiltration has generally not been specified, because of the special techniques required to distinguish between these cell types in tissue sections. When appropriate fixation was used, basophils were found in many DCH reactions, including responses to viruses and contact allergens (5,6). In guinea pigs, basophils comprised as much as 60% of the infiltrating leukocytes in the superficial dermis of DCH reactions (7).

If MCP-1 accounts for the leukocyte infiltration of DCH reactions, it should fulfil the criteria outlined in Table 3.

The number of lymphocytes that react with a specific antigen is too low to generate an effective concentration of MCP-1. MCP-1 must be secreted by resident tissue leukocytes, in response to the very low concentrations of a mediator released by the antigen-reactive lymphocyte. From counts of lymphocytes per mm^2 of

Table 3. Does MCP-1 Mediate DCH Cellular Infiltration?

Interaction of antigen with sensitized lymphocyte should lead to secretion of MCP-1 by unsensitized lymphocytes.
MCP-1 should attract monocytes, basophils, possibly lymphocytes, but not neutrophils.
It should be at the site for at least 24 hrs
Antigen stimulus should not lead to production of NAP-1 or other neutrophil attractants

dermis in sections of normal human skin (8), we estimate that tissue concentration of resident lymphocytes is of the order of 2 to 5×10^6 per ml. Based on this value, and our measurements of MCP-1 in 24 hr culture fluids of PHA-stimulated human lymphocytes, resident tissue lymphocytes could generate an MCP-1 concentration of 10^{-7}M (minus losses due to diffusion out of the region). As to the second criterion in Table 3, it is possible that lymphocyte accumulation in the lesion could be due to replication; whereas the accumulation of monocytes and basophils, which are end-stage cells, must occur by infiltration. The accumulation of basophils illustrates the importance of vasodilatation in DCH. Because of the low concentration of basophils in the circulation, the large numbers of basophils observed in DCH can occur only if large volumes of blood perfuse the reaction site. Since the number of basophils in 24 hr DCH lesions in guinea pigs was equal to the number in the total circulating blood volume (7), it follows that over a period of 24 hrs the amount of blood perfusing the local site could approach in order of magnitude the total blood volume. The role of vasodilatation in attractant-induced leukocyte accumulation has been studied by Williams and Jose (9). The final point in Table 3 relates to the fact that the inability of MCP-1 to attract neutrophils makes it an appropriate mediator of DCH. But in view of the absence of neutrophils at the reaction site, the stimuli for MCP-1 production in DCH should not cause elaboration of NAP-1 or other neutrophil attractants, or else be associated with factors that eliminate NAP-1 activity.

MEASUREMENT OF MCP-1 SECRETION

The simplest way to determine if antigen-stimulated leukocyte cultures produce MCP-1 would be to measure MCP-1 concentrations in culture fluid with a sandwich ELISA. Although we have a polyclonal rabbit anti-MCP-1 antibody, we have been unable to produce a mouse monoclonal anti-MCP-1, and therefore measurement of MCP-1 has been confined to a direct ELISA (step 1: add MCP-1 to reaction well; step 2: block with albumin; step 3: add rabbit anti-MCP-1; step 4: add enzyme-linked goat anti-rabbit IgG; step 5: add substrate). The direct ELISA precludes measurement of MCP-1 in culture fluids, since the presence of other protein (even in the absence of FCS) blocks binding sites in the reaction well. Thus, separation of MCP-1 from culture fluid protein must precede the

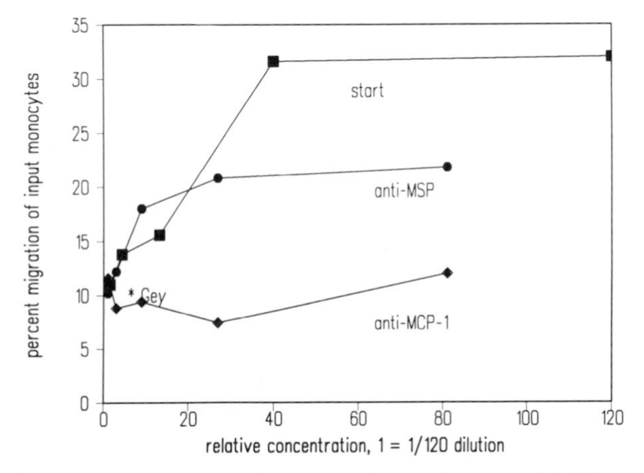

Figure 1. Absorption of chemotactic activity by anti-MCP-1.

direct ELISA. We used an anti-MCP-1 column for separation, measuring loss of chemotactic activity in the pass-through effluent, and MCP-1 by direct ELISA in the protein dissociated from antibody by pH 2.5 glycine buffer. The results of such an experiment with mixtures of elutriated human blood monocytes and lymphocytes stimulated with PHA are shown in Figure 1. The line labeled "start" shows chemotactic activity in 24 hr culture fluids of a PHA-stimulated lymphocyte-monocyte mixture at a ratio of 20:1 ($5x10^6$ lymphocytes per ml). Monocyte chemotactic activity was removed by the anti-MCP-1 column, and partially removed by a column of antibody directed against an unrelated protein (anti-MSP). The presence of MCP-1 in the culture fluid was confirmed by direct ELISA of the glycine eluate; based on this measurement, the minimum MCP-1 concentration in the culture fluid was $4x10^{-8}M$. In contrast, the concentration of MCP-1 in a PHA-stimulated culture of $5x10^6$ monocytes with $.25x10^6$ lymphocytes was only $1.7x10^{-9}M$. This suggests that the producing cells were lymphocytes.

Table 4 shows results of our one experiment to date on antigen stimulation of MCP-1 secretion by human blood mononuclear leukocytes. Although we found no monocyte chemotactic activity (MCA) in 7-day tetanus-stimulated cultures, diphtheria antigen induced significant MCA. By direct ELISA of immunoaffinity column purified MCP-1 from cultures stimulated with 32 LF/ml of diphtheria toxin, the minimum concentration of MCP-1 was $4x10^{-8}M$. This preliminary experiment requires confirmation, as well as determination of the time course of production. If our speculations about stimulation of unsensitized lymphocytes are correct, we might find MCP-1 at 24 hrs.

Table 4. Chemotactic activity of culture fluids of antigen-stimulated mononuclear leukocytes

		migrated monocytes, %
Medium control		7
No antigen		10
PHA		40
Tetanus	8 LF/ml	9
Diphtheria	8 LF/ml	11
	16 LF/ml	17
	32 LF/ml	32
Medium +	32 LF/ml	9

RESPONSE OF BASOPHILS TO MCP-1

We have succeeded in obtaining highly enriched preparations of human blood basophils by a 15 min centrifugation on Percoll of precisely formulated density (10). A typical preparation comprises about 20% basophils (depending on blood basophil concentration of the donor), the other cells being mostly lymphocytes. Using 30,000 basophils per chemotaxis well, we obtain enough cells from 50 ml of blood to fill the 48 wells of the multiwell chemotaxis chamber. Optimal responsiveness and low random migration have been obtained with basophils suspended in Gey's balanced salt solution containing 0.1% low endotoxin bovine albumin (Miles). Figure 2 shows that MCP-1 is an attractant for human basophils; potency is comparable to that of NAP-1, and efficacy (percentage of input cells migrating at optimal concentration) is somewhat higher. In potency and efficacy for basophils, MCP-1 is comparable to monocytes. Since the basophil preparation comprises about 80% lymphocytes, it should be appropriate for measuring lymphocyte responses to

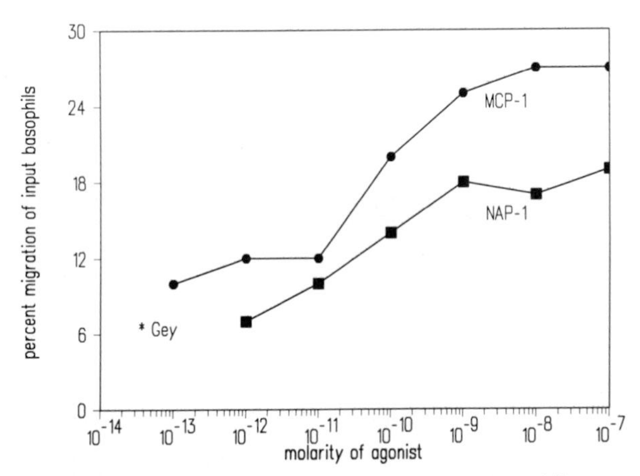

Figure 2. Chemotactic response of human basophils to MCP-1 and NAP-1.

MCP-1. Although we have observed lymphocytes adherent to the MCP-1 attractant side of the polycarbonate membrane used in the chemotaxis assay, the numbers have not exceeded the random migration value. The effect of MCP-1 on lymphocyte migration should be evaluated in cellulosic filter chemotaxis assay (11).

IL-3 has been reported to affect various responses of human basophils. A brief pre-incubation with IL-3 causes LTC_4 secretion and enhanced histamine release in response to C5a (12), and histamine release in response to NAP-1 (13). Figure 3 shows that IL-3 inhibited the efficacy (but not potency) of the MCP-1 attractant response. Partial inhibition of the response to NAP-1 and C5a was also observed. This was not associated with any alteration in basophil morphology as evaluated by light microscopy.

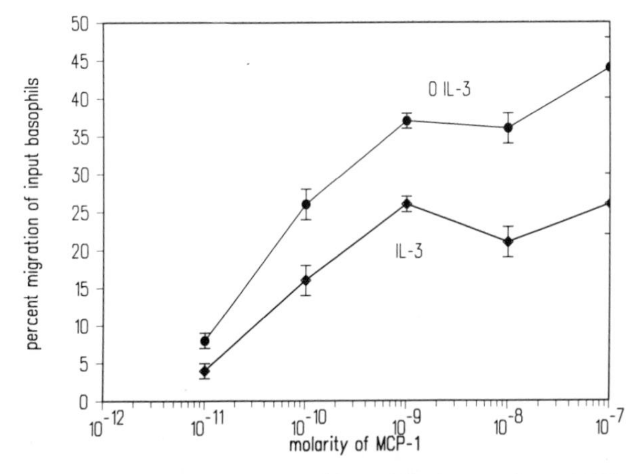

Figure 3. Inhibition by IL-3 of basophil chemotaxis to MCP-1

Table 5. Stimulus-induced activation of MCP-1 and NAP-1 gene expression in different cells.*

Cell Type	Stimuli inducing gene expression for	
	MCP-1	NAP-1
Lymphocyte	PHA, IL-1, LPS	PHA, Con-A
Monocyte	? stimulus	IL-1, TNF, LPS
Macrophage	LPS, Listeria, IFN-γ	LPS, TNF
Fibroblast	PDGF, IL-1, TNF	IL-1, TNF, Rubella
Endothelium	IL-1, TNF, LPS	IL-1, TNF, LPS
Smooth muscle	? stimulus	
Keratinocytes		IL-1, not TNF
Epithelial cells		IL-1, TNF, not LPS
Hepatocytes		IL-1, TNF, not LPS

*The Table is a compilation of data on generation of mRNA or secretion of attractant. For sources, see references 4 and 14.

IS MCP-1 SECRETED WITHOUT NAP-1?

As outlined in Table 3, antigen stimulation in DCH should lead to MCP-1 secretion without secretion of NAP-1. However, Table 5 shows that gene expression for *both* MCP-1 and NAP-1 is found in many cells in response to identical stimuli. What, then, was the selective advantage in evolving leukocyte-specific attractants?

Table 6. A Possible Evolutionary Sequence for MCP-1 and NAP-1

I. Multiple signals cause multiple cells to secrete a non-specific attractant [MN].

II. Gene duplication results in specific M and N attractants, but specific signals for M- and N- gene expression have not evolved.

- IL-1, TNF and LPS all cause endothelial cell MCP-1 and NAP-1 gene expression.
- PHA causes T-lymphocyte MCP-1 and NAP-1 gene expression.

III. M- or N-specific gene expression.

- A new signal evolves
  - PDGF [fibroblast secretion of MCP-1; no NAP-1?]
  - Antigen [T-lymphocyte secretion of MCP-1; no NAP-1?]
- Cell subclasses evolve
  - Bronchoalveolar lavage macrophages [LPS causes NAP-1 secretion; no MCP-1?]
  - Peritoneal macrophages [LPS or IFN-γ cause MCP-1 secretion; no NAP-1?]

A possible evolutionary sequence is outlined in Table 6. We are now evaluating the examples of NAP-1 or MCP-1 specific gene activation listed in section III of Table 6. The evaluation requires measurement of secreted product, and not simply detection of message. Among our first results, we find that in contrast to the NAP-1 concentration of 1000 ng/ml in 24 hr culture fluids of LPS-stimulated human bronchoalveolar lavage macrophages (15), the concentration of MCP-1 is approximately 100-200 ng/ml. The MCP-1 estimate requires confirmation with a sandwich ELISA.

SUMMARY

In this communication, we have asked if MCP-1 is the mediator of cellular infiltration in DCH, outlining the criteria in Table 3. Preliminary data suggest that PHA-stimulated lymphocytes secrete MCP-1, and that MCP-1 can be produced in response to antigen stimulation. MCP-1 attracts monocytes and basophils, but not neutrophils. The question of a lymphocyte response to MCP-1 requires further study. We have emphasized that the discovery of leukocyte-specific NAP-1 and MCP-1 should now be followed by exploration of conditions in which one agonist is secreted without the other. This would be expected, for example, in DCH, which is characterized by mononuclear leukocyte infiltration without neutrophils.

Acknowledgements: We are grateful to Dr. Ilona Sylvester for performing NAP-1 ELISA's and to Dr. Patricia Latham for providing elutriated human monocytes and lymphocytes.

REFERENCES

1. Altman, L.C., R. Snyderman, J.J. Oppenheim, and S.E. Mergenhagen. 1973. A human mononuclear leukocyte chemotactic factor: characterization, specificity and kinetics of production by homologous leukocytes. *J. Immunol.* **110**: 801-810.
2. Altman, L.C. 1978. Chemotactic lymphokines: a review. In *Leukocyte Chemotaxis.* J.I. Gallin and P.G. Quie. eds. Raven Press. New York. p 267.
3. Yoshimura, T., E.A. Robinson, S. Tanaka, E. Appella, and E.J. Leonard. 1989. Purification and amino acid analysis of two human monocyte chemoattractants produced by phytohemagglutinin-stimulated human blood mononuclear leukocytes. *J. Immunol.* **142**: 1956-1962.
4. Leonard, E.J. and T. Yoshimura. 1990. Human monocyte chemoattractant protein-1 (MCP-1). *Immunology Today* **11**: 97-101.
5. Dvorak, H.F. and M.C. Mihm, jr. 1972. Basophil leukocytes in allergic contact dermatitis. *J. Exp Med.* **135**: 235.
6. Askenase, P.W. 1977. Role of basophils, mast cells and vasoamines in hypersensitivity reactions with a delayed time course. *Prog. Allergy* **23**: 199-320.
7. Leonard, E.J., M.A. Lett-Brown, and P.W. Askenase. 1979. Simultaneous generation of tuberculin type and cutaneous basophilic hypersensitivity at separate sites in the guinea pig. *Int. Arch. Allergy Appl. Immuno.* **58**: 460-469.
8. Leonard, E.J., T. Yoshimura, S. Tanaka, and M. Raffeld. 1990. Neutrophil recruitment by intradermally injected neutrophil attractant/activating protein-1 (NAP-1) in human subjects. In *Pathophysiologic and therapeutic roles of cytokines.* Dinarello, C.A., Kluger, M., Powanda, M. and Oppenheim, J.J. eds. Alan Liss, New York. In press.
9. Williams, T.J. and P.J. Jose. 1981. Mediation of increased vascular permeability after complement activation. Histamine-independent action of rabbit C5a. *J. Exp. Med.* **153**: 136-153.
10. El-Naggar, A.L., D.E. van Epps, and R.C. Williams, Jr. 1981. Effect of culturing on the human locomotor response to casein, C5a, and f-Met-Leu-Phe. *Cell Immunol.* **60**: 43.
11. Leonard, E.J. 1987. Two populations of human blood basophils: effect of prednisone on circulating numbers. *J. Allergy Clin. Immunol.* **79**: 775-780.
12. Kurimoto, Y., A.L. de Weck, and C.A. Dahinden. 1989. Interleukin 3-dependent mediator release in basophils triggered by C5a. *J. Exp. Med.* **170**: 467-479.

13. Dahinden, C.A., Y. Kurimoto, A.L. deWeck, I. Lindley, B. Dewald, and M. Baggiolini. 1989. The neutrophil-activating peptide NAF/NAP-1 induces histamine and leukotriene release by interleukin 3-primed basophils. *J. Exp. Med.* 170: 1787-1792.
14. Leonard, E.J. and T. Yoshimura. 1990. Neutrophil attractant/activation protein-1 (NAP-1 [Interleukin-8]). *Am. J. Respir. Cell Mol. Biol.* 2: 479-486.
15. Sylvester, I., J.A. Rankin, T. Yoshimura, S. Tanaka, and E.J. Leonard. 1990. Secretion of neutrophil attractant activation protein by LP-stimulated lung macrophages determined by both enzyme-linked immunosorbent assay and N-terminal sequence analysis. *Am. Rev. Resp. Dis.* **141:** 683-688.

STIMULUS SPECIFIC INDUCTION OF MONOCYTE CHEMOTACTIC PROTEIN-1 (MCP-1) GENE EXPRESSION

Steven L. Kunkel, Theodore Standiford*, Keita Kasahara, and Robert M. Strieter*

Departments of Pathology and *Internal Medicine
Division of Pulmonary and Critical Care
University of Michigan Medical School
Ann Arbor, Michigan 48109-0602

INTRODUCTION

The movement of blood monocytes from the peripheral vasculature into an area of inflammation is an ill defined process. Typically the recruitment of monocytes is associated with a characteristic temporal delay, as these mononuclear phagocytes are histologic markers of chronic cell-mediated responses or delayed type hypersensitivity reactions. While monocytes are typical cells that are found in cell mediated inflammatory lesions, neutrophils are conspicuous by their absence in these lesions. This observation is important for a number of reasons. First, it is apparent that different inflammatory cells are required to perform different functions during the development and maintenance of different inflammatory responses. Second, the low numbers of monocytes in acute inflammatory lesions and high numbers associated with chronic inflammation suggests that specific recruitment signals are expressed to coincide with acute or chronic inflammation. Finally, redundancy exists with regard to chemotactic factors. This is especially true of those chemotactic factors which exhibit activity for more than one population of inflammatory cells. For example, a number of well studied chemotactic factors, such as C5a, leukotriene B4, and fMLP, are potent agents for the movement of both monocytes and neutrophils (Table 1). As mentioned above, monocytes and neutrophils are relatively faithful markers for chronic and acute inflammation, respectively.

This phenomenon becomes even more interesting when one realizes that most chronic responses (monocyte-lymphocyte mediated inflammation) possess an initial neutrophil-mediated acute phase. In addition to specific monocyte and neutrophil chemotactic factors, there must be a unique "switch" mechanism that can down regulate the neutrophil specific or shared chemotactic factors and induce the expression of monocyte specific chemotactic factors. During specific disease states both of the above mechanisms are likely to be operative. In particular, recent scientific advances have been made concerning the identification, isolation, and cloning of specific monocyte chemotactic factors (1,2,3,4). In this manuscript we will describe our recent studies concerning the production and gene expression of a novel monocyte chemotaxin, monocyte chemotactic protein-1 (MCP). Data will

Chemotactic Cytokines, Edited by J. Westwick *et al.*
Plenum Press, New York, 1991

Table 1. List of well studied chemotactic factors with activity for both polymorphonuclear leukocytes and monocytes.

C5a	Leukotriene B_4
fMet Leu Phe	Platelet Activating Factor
Fibronectin	Collagen Products

be presented showing that the induction of steady state mRNA for MCP is dependent upon both the stimulus and cell source under study. In addition, the expression of MCP by many cells is dependent upon cytokine-networking, requiring the prior synthesis of other cytokines in order for MCP expression to occur.

CHARACTERIZATION AND SUPER GENE FAMILY MEMBER

Nearly 20 years have passed since the first immunologic studies were conducted identifying the production of a specific chemotactic factor for mononuclear phagocytic cells (5). These early investigations were important in identifying specific factors that are chemotactic only for monocytes. Recent studies have greatly expanded those original investigations, as numerous group have now identified a specific monocyte chemotactic/activating factor (1,2,3,4,6). Purification studies have now demonstrated that MCP, also known as monocyte chemotactic and activating factor (MCAF), has an apparent molecular weight of approximately 8,500 daltons and consists of 76 amino acids (2). The precursor for MCP is a polypeptide of 99 amino acids (1,2). The mature form of MCP possess 4 cysteines and 13 basic amino acids which potentially confer stability and basic charge properties to this protein (1,2). Monocyte chemotactic protein also possesses a significant degree of homology, at the amino acid level, with IL-8 (neutrophil chemotactic factor) (7). Sequence studies have shown that IL-8 and MCP share 21% homology at the amino acid level (7). Monocyte chemotactic protein belongs to an interesting supergene family of proteins, including a cell competence factor JE (8) and a macrophage inflammatory protein (MIP-1) (9).

Although MCP does belong to a supergene family of different polypeptides, only MCP and JE (the murine equivalent to MCP) possess activity to move peripheral blood monocytes. In order for these chemotactic cytokines to exert a potential biologic effect *in vivo*, a number of important interactions must first occur at the endothelial cell-monocyte level. As shown schematically in Figure 1, the monocyte must first adhere to the endothelial cells, via a mechanism that involves the induction of specific proteins on the endothelium. These endothelial cell-derived proteins are designated as endothelial leukocyte adhesion molecules or intercellular adhesion molecules and operate by binding to a ligand found on monocytes (CD11/CD18 proteins) (10). The induction of these proteins are controlled by proximal mediators which are synthesized early during inflammation, such as tumor necrosis factor (TNF) and interleukin-1 (IL-1) (11). Interestingly, the binding interaction must be strong enough to allow the monocyte to adhere to the endothelium, but still allow the adherence of the endothelial cell/monocyte to be reversible. A similar adherence scenario is known to occur with neutrophils, also mediated by IL-1 and TNF. One major difference lies in the fact that monocytes are not normally marginated in great numbers to the vessel wall as is the case with neutrophils. Thus, the induction of adherence proteins on the endothelium may readily bind neutrophils acutely, but will only bind monocytes

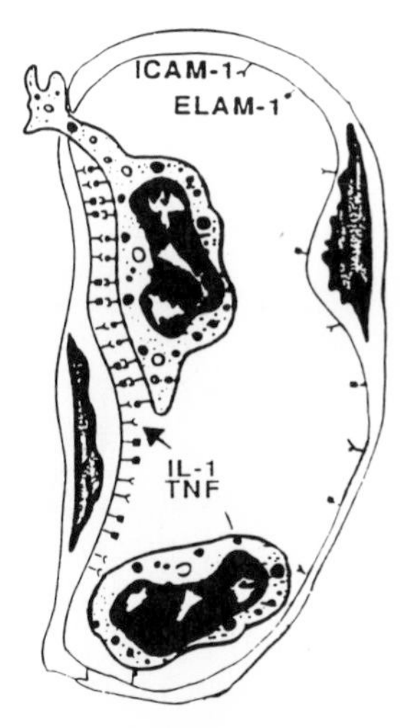

Figure 1. The movement of monocytes from vascular compartment into the interstitium is dependent upon reversible interactions with the endothelium. This interaction involves the induction of specific endothelial cell adherence proteins by TNF and IL-1.

after a lag phase. The lag phase represents the time frame for monocytes in the circulation to come in contact with the endothelial cells.

CYTOKINE-INDUCED MCP GENE EXPRESSION

Experimental studies in our laboratory have assessed MCP gene expression in a number of cell systems (12,13). As shown in Table 2, MCP expression demonstrates both stimulus and cell specificity. While mononuclear phagocytes do not respond to an LPS challenge with the induction of steady state MCP mRNA, endothelial cells express high levels. The endothelial cell appears to be pivotal for the production of MCP as it is the only cell that can respond in a dose and time dependent manner to LPS. An interesting discrepancy exists between the expression of interleukin-8 (neutrophil chemotactic factor) and MCP by human blood monocytes and alveolar macrophages. Stimulation with LPS over a wide time frame resulted in the expression of significant levels of IL-8 mRNA by 2 hours that persisted for 24 hours, while MCP mRNA was not detected over this same time period (Figure 2). The inability of either monocytes or alveolar macrophages

Table 2. The expression of steady state MCP mRNA demonstrates both stimulus specificity and cell source specificity

	Stimulus					
Cell Source	LPS	TNF	IL-1β	IL-1α	IL-4	IL-6
Blood Monocyte	--	--	--	--	--	--
Alveolar Macrophage	--	--	--	--	--	--
Endothelial Cells	+++	++	++++	+++	--	--
Fibroblast	--	++	+++	++	--	--
Epithelial Cell	--	++	+++	++	--	--

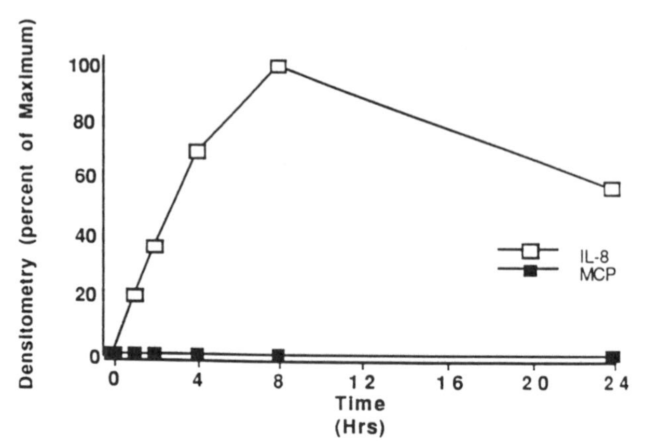

Figure 2. Disparate gene expression of MCP and IL-8 by LPS stimulated alveolar macrophages.

to express MCP mRNA was also observed using IL-1 or TNF as a stimulus (data not shown). While human mononuclear cells do not appear to express MCP mRNA in response to any specific challenge, fibroblasts and epithelial cells demonstrated stimulus specificity. Fibroblasts challenged with IL-1 or TNF, but not LPS, did express MCP mRNA (Figure 3). On the contrary, human endothelial cells could express steady state IL-8 mRNA in response to either LPS or IL-1.

CYTOKINE NETWORKING AND THE INDUCTION OF MCP mRNA

The above data suggest that MCP synthesis may depend upon a cytokine network in order for full expression to occur. The above data also demonstrated that cells once thought of as non-immune bystander cells are actually important "effector" cells with regard to the recruitment of monocytes. The stimulus

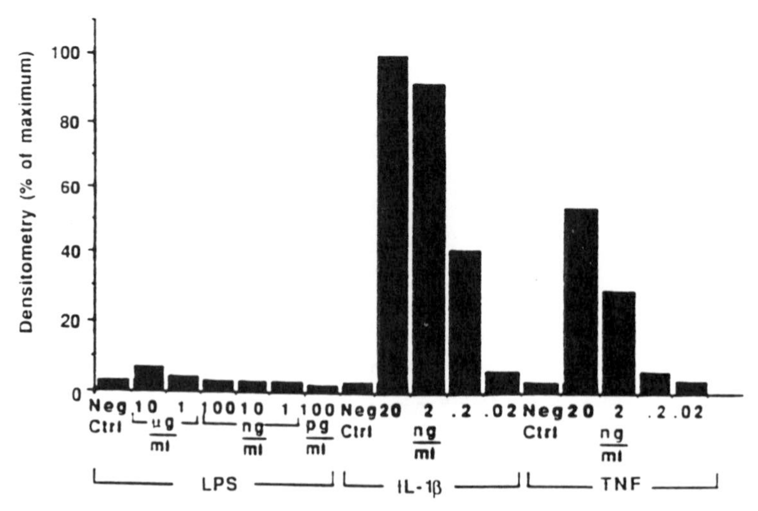

Figure 3. Stimulus specific induction of steady state MCP mRNA by either LPS, IL-1β, or TNF treated fibroblasts.

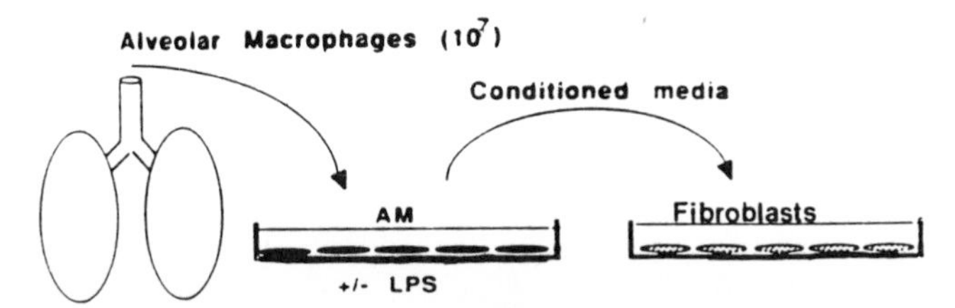

Figure 4. Experimental design for generating LPS stimulated human alveolar macrophage conditioned media used to stimulate human pulmonary fibroblast-derived MCP.

specificity factor may prove also to be important, as the fibroblasts must rely on an initial host response with the production of appropriate cytokines (IL-1 and/or TNF) prior to the production of MCP. Our laboratory has been interested in the phenomenon of cytokine networking that leads to MCP expression.

Figure 4 outlines a specific experiment to truly assess the induction of MCP mRNA via a cytokine network. We have utilized the conditioned media from alveolar macrophages treated with LPS as a stimulus for human lung fibroblast steady state MCP mRNA expression. Since fibroblasts are not responsive to LPS with an increase in MCP mRNA levels, any induction must be due to an agent synthesized by the LPS triggered alveolar macrophages. The conditioned media generated by exposing alveolar macrophages recovered from non-smoking, non-diseased volunteers served as a potent stimulus for the induction of MCP mRNA expression (Figure 5). The laser densitometry of the Northern blots found in Figure 5 also demonstrated that IL-1β was one of the major factors produced by

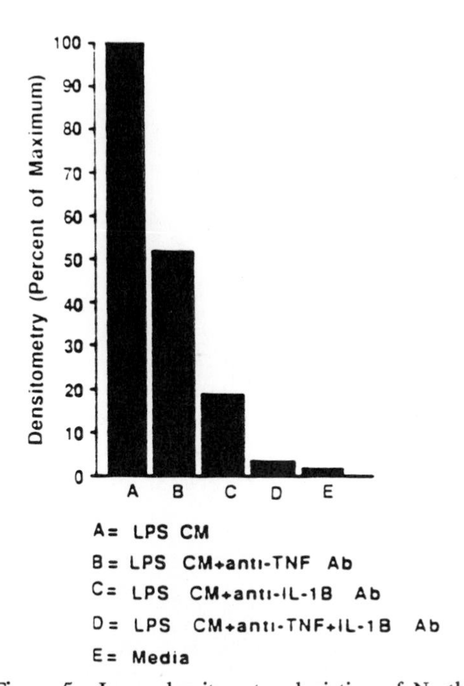

Figure 5. Laser densitometry depiction of Northern blot analysis of alveolar macrophage conditioned media (CM) stimulating pulmonary fibroblast-derived MCP mRNA. CM was used in the presence or absence of specific neutralizing TNF and/or IL-1β antibodies.

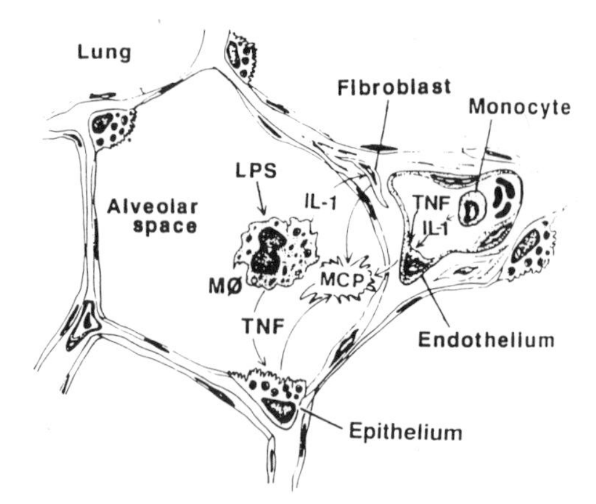

Figure 6. Cytokine networking involving MCP expression by non-immune cells of the pulmonary alveolar space. Alveolar macrophages express IL-1 and/or TNF but not MCP.

the LPS challenged alveolar macrophages. The conditioned media contained stimulatory levels of IL-1β and TNF, as specific neutralizing antibodies to these particular cytokines could reduce the stimulatory activity of the condition media by 82% and 48%, respectively. This study is especially noteworthy as LPS does not serve as a stimulus for MCP mRNA expression by the alveolar macrophages, therefore true networking appears to take place for the production of MCP in the lungs. As shown schematically in Figure 6, the alveolar macrophage can generate TNF and IL-1 when appropriately stimulated, but not MCP. Only when a stimulatory concentration of IL-1 or TNF is generated by the macrophages can MCP be expressed by lung fibroblasts and lung epithelium. This entire process then leads to the elicitation of monocytes into the alveolar interstitium and air space.

SUMMARY

The recruitment of monocytes from the peripheral vasculature to an area of chronic inflammation is a complex phenomenon. This complicated process is likely mediated to a large extent by chemotactic cytokines that are expressed via a cytokine cascade. A number of chronic diseases that are difficult to manage are characterized by a significant infiltrate of monocytes, including sarcoidosis, Wegener's granulomatosis, and tuberculosis. New scientific insight into the mechanisms that lead to early activation events, such as monocyte elicitation, may aid in both better understanding these diverse processes and designing more effective therapies.

Acknowledgements: The authors wish to thank Peggy Otto and Cassandra Narvab for expert secretarial support. This work was supported in part by NIH grants HL31693, HL02401, HL35276, and DK38149. Steven L Kunkel is an established investigator of the American Heart Association.

REFERENCES

1. Yoshimura, T., E.A. Robinson, S. Tanaka, E. Appella, J. Kuratsa, E.J. Leonard. 1989. Purification and amino acid analysis of two human glioma-derived monocyte chemoattractants. *J. Exp. Med.* 169: 1449-1459.

2. Robinson, E.A., T. Yoshimura, E.J. Leonard, S. Tanaka, P.R. Griffin, J. Shabanowitz, D.F. Hunt, E. Appella. 1989. Complete amino acid sequence of a human monocyte chemoattractant, a putative mediator of cellular immune reactions. *Proc. Natl. Acad. Sci. (USA)* 86: 1850-1854.
3. Yoshimura, T., N. Yuhki, S.K. Moore, E. Appella, M.I. Lerman, E.J. Leonard. 1989. Human monocyte chemoattractant protein-1 (MCP). Full length cDNA cloning, expression in mitogen-stimulated blood mononuclear leukocytes, and sequence similarity to mouse competence gene JE. *FEBS* 244: 487-493.
4. Furutani, Y., H. Nomura, M. Notake, Y. Oyamada, T. Fukui, M. Yamada, C.G. Larsen, J.I. Oppenheim, K. Matsushima. 1989. Cloning and sequencing of the cDNA for human monocyte chemotactic and activating factor (MCAF). *Biochem. Biophys. Res. Comm.* 159: 249-253.
5. Ward, P.A., H.G. Remold, J.R. David. 1975. The production of antigen-stimulated lymphocytes of a leukocyte factor distinct from migration inhibitory factor. *Cell. Immunol.* 1: 162-176.
6. Berliner, J.A., M. Territo, L.L. Almada, A. Carter, E. Shafonsky, A.M. Fogelman. 1986. Monocyte chemotactic factor produced by large vessel endothelial cells *in vitro*. *Atherosclerosis* 6: 254-258.
7. Matsushima, K., J.J. Oppenheim. 1989. Interleukin 8 and MCAF: Novel inflammatory cytokines inducible by IL-1 and TNF. *Cytokine* 1: 2-13.
8. Rallins, B.J., E.D. Marrisan, C.d. Stiles. 1988. Cloning and expression of JE, a gene inducible by platelet-derived growth factor and whose product has cytokine-like properties. *Proc. Natl. Acad. Sci. (USA)* 85: 3738-3744.
9. Davatelis, G., P. Tekamp-Olson, S.D. Wolpe, K. Heisen, C. Leudke, C. Grallegas, D. Coit, J. Merryweather, A. Cerami. 1988. Cloning and characterization of a cDNA for murine macrophage inflammatory protein (MIP), a novel monokine with inflammatory and chemokinetic properties. *J. Exp. Med.* 167: 1939-1948.
10. Springer, T.A., M.L. Dustin, T.K. Kishimata, S.D. Martin. 1987. The lymphocyte function associated antigen LFA-1, CD2 and LFA3 molecules. Cell adhesion receptor of the immune system. *Ann. Rev. Immunol.* 5: 223-252.
11. Pohlman, T.H., K.A. Stanness, P.G. Beatty, N.D. Ochs, J.M. Harlan. 1986. An endothelial cell surface factor(s) induced *in vitro* by lipopolysaccharide, interleukin-1 and tumor necrosis factor α increases neutrophil adherence by a CD218-dependent mechanism. *J. Immunol.* 136: 4548-4559.
12. Strieter, R.M., S.W. Chensue, T.J. Standiford, M.A. Baska, H.J. Showell, S.L. Kunkel. 1990. Disparate gene expression of chemotactic cytokines by human mononuclear phagocytes. *Biochem. Biophys. Res. Comm.* 166: 886-891.
13. Strieter, R.M., R. Wiggins, S.H. Phan, B.L. Sharram, H.J. Showell, D.G. Remick, S.W. Chensue, S.L. Kunkel. 1989. Monocyte chemotactic protein gene expression by cytokine-treated human fibroblasts and endothelial cells. *Biochem. Biophys. Res. Comm.* 162: 694-700

GRO: A NOVEL CHEMOTACTIC CYTOKINE

Ruth Sager,[1] Stephen Haskill,[2] Anthony Anisowicz,[1] Douglas Trask,[1] and Marilyn C. Pike[3]

[1]Division of Cancer Genetics
Dana-Farber Cancer Institute, Boston, MA 02115, USA
[2]Department of Obstetrics & Gynecology, Microbiology & Immunology
Lineberger Cancer Research Center, University of North Carolina at Chapel Hill, Chapel Hill, North Carolina 27599-7295, USA
[3]Arthritis Unit, Massachusetts General Hospital
Boston MA 02114, USA

INTRODUCTION

Cytokines are small secreted regulatory peptides that act primarily in a paracrine fashion to provide a signalling system between different cell types. Although cytokine functions were initially associated with myeloid and lymphocytic responses to infection, current studies are demonstrating regulatory functions in growth and differentiation as well. Our studies with the GRO gene represent one of the first to describe a novel cytokine expressed by fibroblasts and epithelial cells as well as cells of the immune system, with regulatory functions both in growth and in the inflammatory response (1, 2, 3).

The GRO gene was discovered in a subtractive hybridization experiment designed to screen for an oncogene or growth factor expressed in Chinese hamster tumorigenic cells but not in their normal counterpart. The gene was recovered as an mRNA over-expressed constitutively in the tumorigenic cells, but transcribed in normal cells only during growth stimulation, e.g. following serum-stimulation of starved cells. This behavior, typical of an early growth-response gene, led to the designation of GRO (1). However, the DNA sequence was related to a series of cytokines including platelet factor 4, platelet basic protein and its split products CTAP III and β-thromboglobulin. These initial findings, suggestive of a growth factor on the one hand and of a cytokine on the other, provided the impetus for our further investigations (reviewed in refs. 2, 3). The GRO gene was shown independently to encode a protein with melanoma growth stimulating activity (MGSA) (4).

This report will summarize recent studies from our laboratory that have shed new light on mechanisms responsible for the diversity of functions of GRO that are now unfolding.

THREE GRO GENES

A human GRO gene was initially recovered by using the hamster cDNA as probe for a cDNA library from the T24 bladder carcinoma cell line, which

Chemotactic Cytokines, Edited by J. Westwick *et al.*
Plenum Press, New York, 1991

constitutively over-expressed GRO mRNA (1). This gene, as well as its homolog encoding the melanoma growth stimulating activity (MGSA) protein, were mapped to chromosome 4q21 (2, 4). Current evidence for the presence of three closely linked but sequentially and functionally independent GRO genes (5) is based on three lines of evidence. (i) Three related but structurally distinct cDNAs were found by subtractive hybridization in activated human monocytes but not in their pre-induced state; (ii) three genes were identified in Southern blots; and (iii) genomic cloning revealed the presence of three distinct GRO genes. The evidence has been recently reported (5) and will be summarized here.

Adherence of monocytes to plastic or to components of the extracellular membrane, or activation by LPS, induces expression of cytokines not expressed in their unstimulated counterparts (6). These cytokines include three closely related GRO genes: the gene that was originally described (1) now called GROα, as well as GROβ and GROγ, which share 90% and 86% identity respectively, with GROα, at the amino acid level (5). The three genes also differ in the 5' and 3' untranslated regions, and the upstream (5') regulatory regions. Since the GROα probe recognizes all three genes, and only one chromosomal locus hybridizes with this probe by hybridization *in situ* (2, 4), it is likely that the three genes arose as tandem repeats and remained closely linked during evolution.

Further evidence for the presence of three distinct genes in the genome comes from Southern hybridizations with the GROα probe, and also with the three gene-specific oligonucleotides. Using the enzymes XbaI and EcoRI to cleave genomic DNA, several hybridizing bands were identified, representing restriction fragments common to the three genes, as well as specific bands characteristic of each gene. Identification of these bands clarified the existence of three distinct genes in the genome (5).

The inferred amino acid sequences of the three GRO genes were compared by computer analysis of their predicted secondary structure. All but two of the amino acid substitutions are conservative in terms of size and hydrophobicity, and in not altering the secondary structural predictions. However there are two proline differences between GROα and GROβ, and one between α and γ. The substitution of proline at amino acid 36 in GROβ for serine in GROα has no apparent effect on structure. The substitution of leucine at amino acid 54 in GROβ and GRO for proline in GROα has a potentially dramatic effect on alpha-helix, beta-sheet, and beta-turn predicted conformation in the secreted protein. This substitution could affect receptor recognition on target cells. The possible influence of additional substitutions, particularly in the carboxy terminus are difficult to predict.

Differential expression of the three GRO genes was found in a study of induction of various cell types (5). In adherent monocytes, where the three mRNAs were initially detected, it was shown that LPS-stimulated monocytes expressed all three genes, whereas non-adhered monocytes stimulated with PMA expressed only GROβ and GROγ. Neutrophils adhered to fibronectin expressed high levels of GROα whereas adherent lymphocytes expressed very low levels of GROα. The effects of induction with IL-1 or TNF were examined, since prior studies had shown that these cytokines induce high levels of GRO mRNA (2). We found that both GROα and GROγ mRNAs were induced to high steady state levels by IL-1 and TNF in human fibroblasts, mammary epithelial cells and umbilical vein endothelial cells, whereas GROβ steady state mRNA levels were very low. The low steady state level of GROβ may result from low transcriptional initiation or high degradation rates, and may also be affected by weak oligonucleotide hybridization to RNA, although the same probe does hybridize strongly to DNA.

Two tumors have been examined. In one, a fresh colon carcinoma explant, only GROγ was expressed, whereas in the bladder carcinoma cell line, T24, only GROα was detected. In previous studies, it was reported that the majority of tumor cell lines examined did not express GRO mRNA at all, whereas others expressed a high level (2). Thus the GRO genes are differentially expressed in different normal cell types and their regulation is aberrant in tumor cells. It will be important to find out whether GRO expression plays a significant role in tumor growth progression.

REGULATION BY NFkB

The upstream region of all three GRO genes contains an NFkB site (GGGAATTTCC) which corresponds to the consensus sequence recognized by the NFkB DNA-binding protein (7). Despite the divergence of sequences in the upstream region of the three genes, the NFkB site is located in a conserved domain. The sequence from the transcription start site is 80-85% conserved among the three GRO genes up to position -105, and the NFkB site is at -60 to -70 in GROβ and at approximately the same position in GROα and GROγ. There is very low conservation from -105 to -145, and then in a domain of about 90 bases, the proximal part is 62% conserved and the distal part is about 83% conserved. There is very little sequence conservation further 5' of -235 (8). The involvement of the NFkB site in the induction of GRO mRNA by pre-treatment of cells with either IL-1 or TNF was shown initially by CAT-analysis. An EcoRI/EspI fragment (-2530 to +70) from the GROβ genomic clone was cloned 5' to the CAT gene in the promotorless CAT vector pKT and a series of nested 5' deletion mutants was constructed using exonuclease III. These constructs were transiently transfected into HeLa cells and the response to IL-1 or TNF was assayed by quantitation of acetylated and non-acetylated chloramphenicol. Sequences from -77 to +70 responded to induction by both IL-1 and TNF as did longer 5' fragments, while sequences from -35 to +70 did not respond. Thus sequences between -77 to -35 were shown to be essential for IL-1 and TNF induction. This region includes the NFkB site at -70 to -60.

In gel retardation experiments, using small fragments from the 5' region, strong binding was detected in extracts of IL-1 or TNF pretreated human fibroblasts incubated with a labelled fragment containing sequences from -77 to +70. Further evidence that the NFkB site was indeed the target of protein-DNA binding came from competitive inhibition studies. Competitive inhibition was observed with a wildtype oligomer of the NFkB sequence but not with a mutant oligomer, previously shown to have diminished NFkB binding capacity (9). Analogous gel retardation experiments were carried out with 5' sequences from each of the three GRO genes with very similar results.

Several cell types have shown NFkB mediated binding and transcriptional response of genes encoding cytokines or other immune responsive proteins. Our studies demonstrate that GRO is a member of this class of genes, and also that similar responsiveness mediated by NFkB is expressed by all three GRO genes.

GRO AS A POTENT NEUTROPHIL CHEMOATTRACTANT

The recombinant GRO protein has been produced from cos cells in this laboratory, purified and identified by its inferred molecular weight (6kD), and in Western blots using a polyclonal antibody produced against a C-terminal oligonucleotide (Trask and Sager, unpublished). The same protein has also been produced as a recombinant product in yeast by Dr. Patricia Beckman (Immunex, Seattle, WA). The two proteins are as yet indistinguishable in functional tests of chemoattraction and as inflammatory agents.

The rGRO protein was found to be chemotactic for human neutrophils (PMNs) assessed by measuring the migration of cells through 3.0 μm polycarbonate filters in 48-well microchambers. rGRO induced a significant migration of PMNs in concentrations ranging from 0.05 nM to 5 nM, with peak migration occurring at 0.5 nM. In four separate experiments, maximal cell migration in response to rGRO occurred at a mean concentration of 0.7 ± 0.2 nM (± S.D.) and an EC_{50} of 0.07 nM. As further evidence of its chemoattractant potency, the total number of cells migrating in response to rGRO ranged from 53 to 83% of the number migrating in response to another chemoattractant, f-Met-Leu-Phe. Checkerboard analysis used to distinguish between chemotaxis and chemokinesis (12), demonstrated that GRO is not a chemokinetic agent, but that its motility-inducing effect on neutrophils is chemoattractant: a positive gradient of GRO protein was necessary to enhance migration. Unlike neutrophils, monocytes did not respond chemotactically to GRO concentrations between 0.01 and 10 nM (11).

The chemotactic role of GRO is consistent with other evidence that GRO is a cytokine, involved in the inflammatory response. The protein was shown to induce an acute inflammatory reaction when injected into the footpads of endotoxin resistant C3H mice (10). GRO differs functionally from the related cytokine NAP-1/IL-8 (13) in not inducing superoxide production by neutrophils. NAP-1/IL-8 and GRO share 43 amino acid identity and strong structural similarity as shown by comparison of their sequences and hydrophobicity plots.

Chemotactic and inflammatory activities similar to those of GRO have been described for the mouse gene product MIP-2 (14). Considerable sequence relatedness to GRO has been described for the mouse gene KC (15). As yet, GRO itself has been the closest human relative of both MIP-2 and KC that has been identified. As described at this meeting, the human counterparts of mouse MIP-2, called human MIP-2α and MIP-2β have sequences identical with GROβ and GROγ respectively at the amino acid level (16).

Thus, among the human C-X-C cytokine sub family genes, GRO and NAP-1/IL-8 appear to be the most closely related. As yet their principal functional difference appears to lie in the ability of IL-8 to induce neutrophil, monocyte and lymphocyte chemotaxis whereas GRO chemotaxis is limited to neutrophils. However, in view of the evolutionary conservation of both cytokines as well as the diverse regulation of expression of the three GRO genes, it seems evident that current research has only begun to unravel the functions of these genes and the significance of their finely tuned expression. Studies in progress to identify and clone the GRO receptor should facilitate future GRO research. Of particular interest in our laboratory is the role of GRO and related cytokines in human cancer, and their potential use as anti-cancer agents.

REFERENCES

1. Anisowicz, A., L. Bardwell, and R. Sager. 1987. Constitutive over-expression of a novel growth-regulated gene in transformed Chinese hamster and human cells. *Proc. Natl. Acad. Sci. USA* **84**: 7188-7192.
2. Anisowicz, A., D. Zajchowski, G. Stenman, and R. Sager. 1988. Functional diversity of GRO gene expression in human fibroblasts and mammary epithelial cells. *Proc. Natl. Acad. Sci. USA* **85**: 9645-9649.
3. Sager, R. 1990. GRO as a cytokine. In: *Molecular and Cellular Biology of Cytokines.* M.C. Powanda, J.J. Oppenheim, M.J. Kluger, and C.A. Dinnarello, eds. Alan Liss, New York.
4. Richmond, A., E. Balentien, H.G. Thomas, J. Spiess, R. Bordoni, U. Francke, and R. Derynck. 1988. Molecular characterization and chromosomal mapping of melanoma growth stimulating activity, a growth factor structurally related to β-thromboglobulin. *EMBO J.* 7: 2025-2033.
5. Haskill, S., A. Peace, J. Morris, S.A. Sporn, A. Anisowicz, S.W. Lee, T. Smith, G. Martin, P. Ralph, and R. Sager. 1990. Identification of three related human GRO genes encoding cytokine functions. *Proc. Natl. Acad. Sci. USA* (in press).

6. Sporn, S.A., D.F. Eierman, C.E. Johnson, G. Martin, M. Ladner, and S. Haskill. 1990. Monocyte adherence to matrix components results in selective induction of novel genes sharing homology with mediators of inflammation and tissue repair. *J. Immunol.* **144**: 4434-4441.
7. Lenardo, M., and D. Baltimore. 1989. NFkB: a pleiotropic mediator of inducible and tissue-specific gene control. *Cell* **58**: 227-229.
8. Anisowicz, A., M. Messineo, S. Lee, and R. Sager. 1990. NFkB mediates IL-1/TNF induction of GRO in human fibroblasts. (in preparation).
9. Lenardo, M., C. Fan, T. Maniatis, and D. Baltimore. 1989. The involvement of NFkB in γ-interferon gene regulation reveals its roles as widely inducible mediator of signal transduction. *Cell* **58**: 287-294.
10. Pike, M.C., D. Trask, and R. Sager. 1990. Recombinant GRO gene product ischemotactic for human neutrophils. *Clin. Res.* **38**: 479a.
11. Pike, M., D. Trask, and R. Sager. 1990. Recombinant human GRO is chemotactic for human polymorphonuclear leukocytes. (submitted).
12. Zigmond, S.H., and J.G. Hirsch. 1973. Leukocyte locomotion and chemotaxis - methods for evaluation and demonstration of a cell-derived chemotactic factor. *J. Exp. Med.* **137**: 387-392.
13. Matsushima, K., K. Morishita, T. Yoshimura, S. Lavu, Y. Kobayashi, W. Lew, E. Appella, H.F. Kung, E.J. Leonard, and J.J.Oppenheim. 1988. Molecular cloning of a human monocyte-derived neutrophil chemotactic factor (MDNCF) and the induction of MDNCF mRNA by interleukin 1 and tumor necrosis factor. *J. Exp. Med.* **167**: 1883-1893.
14. Wolpe, S.D., B. Sherry, D. Juers, G. DaVatelis, R.W. Yurt, andA. Cerami. 1989. Identification and characterization of macrophage inflammatory protein 2. *Proc. Natl. Acad. Sci. USA* **86**: 612-616.
15. Oquendo, P., J. Alberta, D. Wen, J.L. Graycar, R. Derynck, and C.D. Stiles.1989. The platelet-derived growth factor-inducible KC gene encodes a secretory protein related to platelet α-granule proteins. *J. Biol. Chem.* **264**: 4133-4137.
16. Tekamp-Olson, P., C. Gallegos, D. Bauer, J. McClain, B. Sherry, M. Fabre, S. van Deventer, and A. Cerami. 1990. Cloning and characterization of cDNAs for murine MIP-2 and human MIP-2 homologs. (this book, abstract).

PDGF AND THE SMALL INDUCIBLE GENE (SIG) FAMILY: ROLES IN THE INFLAMMATORY RESPONSE

Rodney S. Kawahara[1], Zheng-Wen Deng[1], and Thomas F. Deuel[1,2]

Departments of [1]Medicine and [2]Biochemistry/Molecular Biophysics
Jewish Hospital at Washington University Medical Center
216 South Kingshighway, St. Louis, Missouri 63110, USA

INTRODUCTION

Inflammation and tissue repair involve a series of highly regulated and coordinated reactions that are beginning to be recognized and understood. Molecules ordinarily within intracellular compartments such as the platelet α-granule are released and appear to interact with target cells to initiate cell migration and division (1). The release of these factors and their interaction with cells stimulates a second wave of events through the transcriptional activation of genes which encode proteins with additional signalling roles (2,3). Inflammatory cells have been identified in close proximity to platelets in models of inflammation and immune complex disease and in lesions of atherosclerosis (4–8). These findings suggest close interactions and exchange of signals between platelets and inflammatory cells that mutually stimulate the release of these signalling molecules and initiate subsequent transcriptional events for the synthesis of additional mediators of inflammation and wound repair. Platelets may therefore play a central role in inflammation and tissue repair.

EARLY MEDIATORS OF THE INFLAMMATORY RESPONSE

Platelet factor 4 (PF4) is released from platelets when platelets are activated by thrombin or other agonists at sites of inflammation and vascular wall injury. PF4 was the first platelet α-granule protein that was shown to be chemotactic for inflammatory cells (9). Chemotaxis is stimulated by PF4 at concentrations less than that found in human serum (10). PF4 is also strongly chemotactic for fibroblasts and is at least ten times more potent for fibroblast chemotaxis as compared to concentrations of PF4 required for maximum chemotaxis of inflammatory cells (11). This might suggest that PF4 preferentially stimulates fibroblast migration into wound healing sites. Platelet-derived growth factor (PDGF) is also strongly chemotactic for human monocytes and neutrophils with maximum activity at PDGF concentrations well below those found in serum (12). PDGF is also a chemoattractant for fibroblasts and smooth muscle cells (11–15). β-Thromboglobulin (β-TG), another platelet secretory α-granule protein, is highly active as a chemoattractant for fibroblasts (11).

Other early cellular effects important to inflammation and wound repair have been observed to be initiated by platelet α-granule secretory proteins. PDGF

Chemotactic Cytokines, Edited by J. Westwick *et al.*
Plenum Press, New York, 1991

activates human polymorphonuclear leukocytes (16) and monocytes (17). It also stimulates an increase in the release of collagenase by fibroblasts in culture (18), an effect that may be important in tissue repair (19-21). Recently, a peptide derived from the fifth component of complement, collagens and collagen-derived peptides, fibronectin, tropoelastin, and elastin-derived peptides have been shown to have fibroblast chemotactic activity. These early results suggest that platelets have the potential to initiate the complete early events of inflammation through the release of platelet α-granule products from intracellular stores by attracting the cells important in subsequent events leading to inflammation and tissue repair.

EARLY MEDIATORS OF THE INFLAMMATORY RESPONSE ALSO ACTIVATE THE TRANSCRIPTION OF QUIESCENT GENES ENCODING CYTOKINES

In response to PDGF, fibroblasts transcribe and express otherwise quiescent genes. Along with the expression of genes which mediate the intracellular propagation of the mitogenic signal, PDGF induces the expression of genes which encode cytokines that are important for intercellular communication. These cytokines such as JE, KC/gro/MGSA, IL-6, and M-CSF (20-24) are important initiators and regulators of inflammatory cells that serve to coordinate the diverse responses which are designed to combat wound infection and initiate healing. The PDGF early response genes *JE* and *KC* were first isolated by differential colony hybridization (22). These genes subsequently were identified as cytokines which belong to a newly described superfamily of small inducible genes (SIG) (25) which include the α-granule proteins platelet-basic protein, βTG and platelet factor 4.

The murine *JE* gene is composed of three exons and two introns and encodes a 148 amino acid basic (pI = 10.4) polypeptide (24, 25). The rat *JE* gene (26) similarly encodes a 148 amino acid protein and is 82% homologous to the murine *JE* gene. Both rat and murine *JE* gene products contain amino-terminal hydrophobic leader sequences and virtually identical alternating hydrophobic and hydrophilic domains. The single N-linked glycosylation site at position 126 is conserved in both the murine and rat *JE* genes and the intron-exon splice junctions of both genes are identical. The human homolog of the rodent *JE* gene has been identified as being 68% identical with monocyte chemotactic protein-1 (MCP-1) (27). MCP-1 is identical with MCAF (28), MCP (29), HC11 (30), and may be identical with smooth muscle cell-chemotactic factor (SMC-CF) (31). MCP-1 is 49 amino acids shorter at the carboxy-terminus than the rodent *JE* gene product, suggesting either that the carboxy-terminus is not important for a functional JE molecule or that the rodent *JE* gene product may have other species specific functions. MCP-1 and rodent *JE* are induced in fibroblasts by the same stimuli (PDGF, serum, TPA, double stranded RNA, IL-1) and cross hybridize to the same bands on Southern blots (32). Antibodies raised to the murine JE protein cross-reacts with MCP-1.

MCP-1 is chemotactic for monocytes but not for neutrophils (33). Other well characterized chemotactic factors do not show a similar monocyte specificity (34), suggesting that MCP-1 may have a unique role in the inflammatory process. MCP-1/SMC-CF or related molecules appear to be the major chemotactic factors released by a number of tumor cell lines (31).

THE SMALL INDUCIBLE GENE FAMILY

The *JE* gene is a member of a newly recognized family (Figure 1) of small inducible genes (SIG) which include platelet α-granule proteins which are stored and released with PDGF during platelet aggregation (24-63). Other members of

```
               |  EXON 1           |  EXON 2                                   |  EXON 3
mu JE          HVLAQP▽DAVNAPLT    C-C  YSF-TSKMIPMSRLESYKRITSSR   C  PK-EAVV▽FVTKLKREV    C  ADPKKEWVQTYI-KNLD
ra JE          HVLSQP DAVNAPLT    C-C  YSF-TGKMIPMSRLENYKRITSSR   C  PK-EAVV FVTKLKREI    C  ADPNKEWVQKYI-RKLD
hu MCP-1       QGLAQP DAINAPVT    C-C  YNF-TNRKISVQRLASYRRITSSK   C  PK-EAVI FKTIVAKEI    C  ADPKQKWVQDSM-DHLD
hu HC14           AQP■■DSVSIPIT   C-C  FNV-INRKIPIQRLESYTRITNIQ   C  PK-EAVI■■FKTKRGKEV   C  ADPKERWVRDSM-KHLD
hu LD78α       NQFSAS LAADTPTA    C-C  FSY-TSRQIPQNFIADYF-ETSSQ   C  SKP-GVI FLTKRSRQV    C  ADPSEEWVQKYVSDLEL
hu LD78ß       NVLSAP LAADTPTA    C-C  FSY-TSRQIPQNFIADYF-ETAAQ   C  SKP-SVI FLTKRGRQV    C  ADPSEEWVQKYVSDLEL
mu MIP-1α      QVFSAP■■YGADTPTA   C-C  FSY--SRKIPRQFIVDYF-ETSSL   C  SQP-GVI■■FLTKRNRQI   C  ADSKETWVQEYITDLEL
mu MIP-1ß      PGFSAP■■MGSDPPTS   C-C  FSY-TSRQLHRSFVMDYY-ETSSL   C  SKP-AVV■■FLTKRGRQI   C  ANPSEPWVTEYMSDLEL
hu HC21        PALSAP■■MGSDPPTA   C-C  FSY-TARKLPRNFVVDYY-ETSSL   C  SQP-AVV■■FQTKRSKQV   C  ADPSESWVQEYVYDLEL
hu ACT-2       PALSAP■■MGSDPPTA   C-C  FSY-TARKLPRNFVVDYY-ETSSL   C  SQP-AVV■■FQTKRSKQV   C  ADPSESWVQEYVYDLEL
hu H400        PALSAP■■MGSDPPTA   C-C  FSY-T-REASSNFVVDYY-ETSSL   C  SQP-AVV■■FQTKRSKQV   C  ADPSESWVQEYVYDLEL
mu H400        PGFSAP■■MGSDPPTS   C-C  FSY-TSRQLHRSFVMDYY-ETSSL   C  SKP-AVV■■FLTKRGRQI   C  ANPSEPWVTEYMSDLEL
hu RANTES      P-ASAS■■PYSSDTTP   C-C  FAY-IARPLPRAHIKEYF-YTSGK   C  SNP-AVV■■FVTRKNRQV   C  ANPEKKWVREYINSLEM
hu TCA3        QDVDSK■■SMLTVSNS   C-C  LN-TLKKELPLKFIQCYRKMGSS-   C  PDPPAVV■■FRLNKGRES   C  ASTNKTWVQNHLKKVNP

hu IP10        IQ▽GVPLSRT---VR    CTC  ISISNQPVNPRSLEKLEIIPASQF   C  PRVEII▽ATMKKKGEKR    C  LNPESKAIKNLLKAVSK
hu NAP-1/Il-8  CE GAVLPRSAKELR    CQC  IKTYSKPFHPKFIKELRVIESGPH   C  ANTEII VKL-SDGREL    C  LDPKENWVQRVVEKFLK
hu PF4         AS AEAEEDG--DLQ    CLC  VKTTSQ-VRPRHITSLEVIKAGPH   C  PTAQLI ATL-KNGRKI    C  LDLQAPLYKKII-KKLL
hu PF4VAR      AR AEAEEDG--DLQ    CLC  VKTTSQ-VRPRHITSLEVIKAGPH   C  PTAQLI ATL-KNGRKI    C  LDLQALLYKKII-KEHL
ra PF4         TR ASPEESDG-DLS    CVC  VKTSSSRIHLKRITSLEVIKAGPH   C  AVPQLI ATL-KNGSKI    C  LDRQVPLYKKIIKKLLE
bo PF4         PA■■DSEGGEDE-DLQ   CVC  LKTTS-GINPRHISSLEVIGAGTH   C  PSPQLL■■ATK-KTGRKI   C  LDQQRPLYKKIL-KKLL
hu CTAPIII     KE■■ESLDSDLYAELR   CMC  IKTTS-GIHPKNIQSLEVIGKGTH   C  NQVEVI■■ATL-KDGRKI   C  LDPDAPRIKKIVQKKLA
hu GRO         GR■■RAAGASVATELR   CQC  LQTL-QGIHPKNIQSVNVKSPGPH   C  AQTEVI■■ATL-KNGRKA   C  LNPASPIVKKIIEKMLN
ha GRO         SR■■LATGAPVANELR   CQC  LQTMT-GVHLKNIQSLKVTPPGPH   C  TQTEVI■■ATL-KNGQEA   C  LNPEAPMVQKIVQKMLK
mu KC          SR■■LATGAPIANELR   CQC  LQTMA-GIHLKNIQSLKVLPSGPH   C  TQTEVI■■AYL-KNGREA   C  LDPEAPLVQKIVQKMLK
ra CINC                APVANELR   CQC  LQTVA-GIHFKNIQSLKVMPPGPH   C  TQTEVI■■ATL-KNGREA   C  LDPEAPMVQKIVQKMLK
ch 9E3         SQ■■GRTLVKMGNELR   CQC  ISTHSKFIHPKSIQDVKLTPSGPH   C  KNVEII■■ATL-KDGREV   C  LDPTAPWVQLIV-KALM
mu MIP-2             AVVAS--ELR   CQC  LKTLPR-VDFKNIQSLSVTPPG
```

Figure 1. The small inducible gene (SIG) family. The intron–exon splice junctions are denoted by ▽ for family members where data is available or predicted ■ based on genetic homology. Gaps in the amino acid sequence alignment are denoted by (-).

the family are inducible by various stimuli (Table 1) and many were discovered by differential colony hybridization. SIG family members display spatially conserved cysteine and proline residues. Within the family, there appears to be two subgroups which may represent tandem duplications of related family members at two genetic loci (64,65). One subgroup is recognized by having two adjacent cysteine residues similar to the *JE* gene product while the other subgroup which includes platelet factor 4 has as a double cysteine separated by a single amino acid.

The genomic organization of the SIG family genes sequenced so far suggest that these genes may have evolved from a common ancestor. Although the lengths of the introns may vary, the location of the first and second introns have been conserved. Most SIG family members are composed of two introns and three exons except NAP-1/IL-8 and IP-10. Both NAP-1/IL-8 and IP-10 have an extra intron and a fourth exon which codes for the last few amino acid residues. This is probably due to genetic drift, since both the amino and carboxyl-terminal regions of SIG family genes are diverse. Exon 3 contains a WVQ motif in many but not all SIG family members of both subgroups. The significance of this motif is not currently known.

Table 1 summarizes many of the SIG family encoded proteins and their prototypic inducers. Each of the genes are normally not expressed and can be induced in the presence of the appropriate stimuli. Furthermore, each of the inducers has an important role in cell growth, inflammation, or immune responses and some of the protein products of the SIG family are known to have important similar roles.

Table 1. Prototype inducers which stimulate the transcription of SIG family members

Gene	Inducer	Cell
JE	PDGF	Balb/c 3T3 fibroblast
MCP-1	PDGF	WI-38 fibroblast
HC11	gamma-interferon	U937 macrophage
HC14	gamma-interferon	U937 macrophage
LD78	PHA	Human tonsillar lymphocytes
MIP-1α	LPS	RAW264 macrophage
MIP-1β	LPS	RAW264 macrophage
HC21	anti-$T11_2$ and $T11_3$ antibodies	Human peripheral blood lymphocytes
Act-2	PHA	Human peripheral blood lymphocytes
H400	ConA	$C1Ly-1^+2^-/9$ T cell
RANTES	PHA	4H2 T cell
TCA-3	ConA	C1.Ly-T1 T cell
IP-10	gamma-interferon	U937 macrophage
NAP-1/IL-8	Staph. Enterotoxin A	U937 macrophage
KC/gro/MGSA	PDGF	Balb/c 3T3 fibroblast
9E3	serum	Chicken fibroblast
MIP-2	endotoxin	RAW 264.7 macrophage

REGULATION OF *JE* GENE EXPRESSION BY GLUCOCORTICOIDS

Glucocorticoids are potent anti-inflammatory agents which act in part by inhibiting the production of key cytokines and lymphokines both transcriptionally and translationally. This action effectively disrupts the normal intercellular communication involved in the immune and inflammatory response. The addition of dexamethasone to PDGF and serum stimulated Balb/c 3T3 fibroblasts prevented the normal induction of the *JE* gene by serum and PDGF in a dose dependent manner with the expected appropriate rank order of potency expected for glucocorticoid receptor mediated anti-inflammatory activity. This inhibition was not due to a shift in the time course of induction and non-steroidal anti-inflammatory agents did not show similar activity. Figure 2 summarizes three possible mechanisms which might explain how glucocorticoids might reverse the induction of the *JE* gene by serum and PDGF.

Possible Mechanisms for Glucocorticoid Regulation

Figure 2 summarizes the possible mechanisms for glucocorticoid regulation. One mechanism suggests that glucocorticoids act by interfering with the growth factor initiated signal transduction system. If this were true, the target of interaction would require a common pathway between PDGF, TPA, and double stranded (poly rI:rC) RNA since glucocorticoids can reduce the *JE* induction by these stimuli. In contrast, the induction of the *c-myc* gene was not inhibited by glucocorticoids implying that not all signalling pathways are affected by glucocorticoids. A second alternative hypothesis suggests that the negative regulation of the *JE* gene is mediated by the direct binding of glucocorticoid receptor to glucocorticoid response elements (GRE) located within the *JE* promoter. This GRE is hypothesized to overlap the binding site of a positive transcription factor and thus prevents the interaction of the positive factor with the promotor. The glucocorticoid receptor could alternatively act by increasing the selective degradation of the *JE* mRNA by direct protein interaction. In contrast to the hypotheses which propose direct

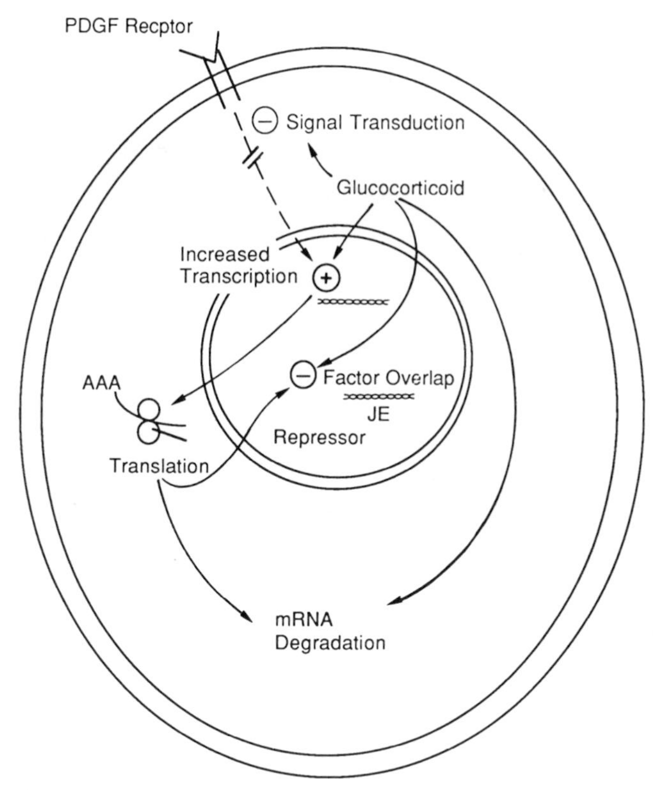

Figure 2. Possible mechanisms of glucocorticoid mediated inhibition of *JE* induction by PDGF and serum.

interaction with the glucocorticoid receptor, a third alternative suggests that the negative regulation of the *JE* gene might be a secondary effect due to the transcriptional induction and expression of a labile repressor for the *JE* gene. Alternatively glucocorticoids may stimulate the induction and expression of specific factors which decrease the stability of the *JE* mRNA.

Direct action of the glucocorticoid receptor (the second hypothesis) may be distinguished from indirect actions of the receptor (the third hypothesis) by the use of protein synthesis inhibitors. Since the third hypothesis requires new protein synthesis but the second hypothesis does not, indirect actions of the glucocorticoid receptor would be predicted to be sensitive to the presence of cycloheximide and puromycin and result in the loss of the glucocorticoid mediated transcriptional inhibition. The addition of protein synthesis inhibitor cycloheximide and puromycin were found to block the negative regulation of the *JE* gene by glucocorticoids suggesting that new protein synthesis was required. Nuclear runoff assays showed that the major mechanism of glucocorticoid inhibition was at the initiation of transcription and the addition of puromycin in the presence of dexamethasone could reverse the glucocorticoid dependent block in transcription. The addition of dexamethasone or puromycin did not significantly alter the rate of *JE* mRNA degradation suggesting that the predominant mechanism of regulation was at the transcriptional level and not because of *JE* message destabilization. Experiments are currently being conducted to identify the promoter control regions which are responsible for the negative regulation as well as the identity of the labile repressor.

Table 2. Summary of the Effect of Various Stimuli on *JE* Gene Induction in Balb/C 3T3 Cells

Inducer	*JE* Induction	*JE* Induction with Dexa-methasone	*JE* Induction with Dexamethasone and Puromycin
Serum	+++	-	+++
PDGF	+++	-	+++
TPA	++	-	ND
poly rI:rc	++++	++	ND
Forskolin	-	-	ND

ND = Not Done

SUMMARY

A growing body of evidence suggests that cells responding to an initial growth, inflammatory or immune signal can respond by inducing the transcription of selective members of the SIG family. This family of related inducible cytokines are involved in the amplification, propagation and coordination of intercellular communication among cell types involved in the immune and inflammatory responses. The principal anti-inflammatory action of glucocorticoids may involve the transcriptional and translational inhibition of cytokines such as JE and other members of the SIG family to effectively disrupt the normal lines of intercellular communication which normally coordinates the immune and inflammatory response. The identification of new members of the family and the discovery of the functions of the known members will lead to a clearer understanding of the complicated processes which lead to normal and pathological immune and inflammatory reponses.

REFERENCES

1. Deuel, T. F. 1987. Polypeptide growth factors: Roles in normal and abnormal cell growth. *Ann. Rev. Cell. Biol.* 3: 443-492.
2. Deuel, T. F., N. J. Silverman and R. S. Kawahara. 1988. Platelet-derived growth factor: a multifunctional regulator of normal and abnormal cell growth. *Biofactors* 1: 213-217.
3. Rollins, B. J. and C. D. Stiles. 1989. Serum-inducible genes. *Advan. Can. Res.* 53: 1-32.
4. Senior, R. M., G. L. Griffin, A. Kimura, J. S. Huang and T. F. Deuel. 1989 Platelet α-granule protein-induced chemotaxis of inflammatory cells and fibroblasts. *Meth. Enzymol.* 169: 233-244.
5. Packham, M. A. and J. F. Mustard 1986. The role of platelets in the development and complications of atherosclerosis. *Sem. Hematol.* 23: 8-26.
6. Devreotes, P. N., and S. H. Zigmond. 1988. Chemotaxis in eukaryotic cells: a focus on leukocytes and dictyostelium. *Ann. Rev. Cell. Biol.* 4: 649-686.
7. Wilkinson, P. C., 1982 *Chemotaxis and Inflammation*, Churchill Livingstone, New York.
8. Nachman, R. L. and B. Weksler. 1980. *The Cell Biology of Inflammation* (G. Weissmann, ed.), p. 145. Elsevier/North Holland, New York.
9. Deuel, T. F., R. M. Senior, D. Chang, G. L. Griffin, R. L. Heinrikson and E. T. Kaiser. 1981. Platelet factor 4 is chemotactic for neutrophils and monocytes. *Proc. Natl. Acad. Sci. USA* 78: 4584-4587.
10. Osterman, D. G., G. L. Griffin, R. M. Senior, E. T. Kaiser, and T. F. Deuel. 1982. The carboxyl-terminal tridecapeptide of platelet factor 4 is a potent chemotactic agent for monocytes. *Biochem. Biophys. Res. Comm.* 107: 130-135.
11. Senior, R. M., G. L. Griffin, J. S. Huang, D. A. Walz, and T. F. Deuel. 1983. Chemotactic activity of platelet alpha granule proteins for fibroblasts. *J. Cell. Biol.* 96: 382-385.
12. Deuel, T. F., R. M. Senior, J. S. Huang, and G. L. Griffin. 1982. Chemotaxis of monocytes and neutrophils to platelet-derived growth factor. *J. Clin. Invest.* 89:1046-1049.

13. Grotendorst, G. R., H. E. J. Seppa, H. K. Kleinman, and G. R. Martin. 1981. Attachment of smooth muscle cells to collagen and their migration toward platelet-derived growth factor. *Proc. Natl. Acad. Sci. USA* 78: 3669-3672.
14. Seppa, H., G. Grotendorst, S. Seppa, E. Schiffmann, and G. R. Martin. 1982. Platelet-derived growth factor is chemotactic for fibroblasts. *J. Cell. Biol.* 93: 584-588.
15. Siegbahn, A., A. Hammacher, B. Westermark and C.-H. Heldin. 1990. Differential effects of the various isoforms of platelet-derived growth factor on chemotaxis of fibroblasts, monocytes and granulocytes. *J. Clin. Invest.* 85: 916-920.
16. Tzeng, D. Y., T. F. Deuel, J. S. Huang, R. M. Senior, L. A. Boxer, and R.L. Baehner. 1984. Platelet-derived growth factor promotes polymorphonuclear leukocyte activation. *Blood* 64: 1123-1128.
17. Tzeng, D. Y., T. F. Deuel, J. S. Huang, and R. L. Baehner. 1985. Platelet-derived growth factor promotes human peripheral monocyte activation. *Blood* 66: 179-183.
18. Bauer, E. A., T. W. Cooper, J. S. Huang, J. Altman and T. F. Deuel. 1985. Stimulation of *in vitro* human skin collagenase expression by platelet-derived growth factor. *Proc. Natl. Acad. Sci. USA* 82: 4132-4136.
19. Gross, J. 1976 *Biochemistry of Collagen* (G. N. Ramachandran and A. H. Reddi, eds), p. 275. Plenum, New York.
20. Pierce, G. F., T. A. Mustoe, J. Lingelbach, V. R. Masakowski, P. Gramates and T. F. Deuel. 1989. Transforming growth factor β reverses the glucocorticoid-induced wound-healing deficit in rats: Possible regulation in macrophages by platelet-derived growth factor. *Proc. Natl. Acad. Sci. USA* 86: 2229-2233.
21. Pierce, G. F., T. A. Mustoe, J. Lingelbach, V. R. Masakowski, G. L. Griffin, R. M. Senior and T. F. Deuel. 1989. Platelet-derived growth factor and transforming growth factor-β enhance tissue repair activities by unique mechanisms. *J. Cell. Biol.* 109: 429
22. Cochran, B. H., A. C. Reffel, and C. D. Stiles. 1983. Molecular cloning of gene sequences regulated by platelet-derived growth factor. *Cell* 33: 939-947.
23. Kohase, M., L. T. May, I. Tamm, J. Vilcek, and P. B. Sehgal. 1987. A cytokine network in human diploid fibroblasts: interaction β-interferons, tumor necrosis factor, platelet-derived growth factor, and interleukin-1. *Mol. Cell. Biol.* 7: 273-280.
24. Rollins, B. J., E. D. Morrison, and C. D. Stiles. 1988. Cloning and expression of *JE*, a gene inducible by platelet-derived growth factor and whose product has cytokine-like properties. *Proc. Natl. Acad. Sci. USA* 85: 3738-3742.
25. Kawahara, R. S. and T. F. Deuel. 1989. Platelet-derived growth factor inducible gene *JE* is a member of a family of small inducible genes related to platelet factor 4. *J. Biol. Chem.* 264: 679-682.
26. Timmer, H. T. M., G. J. Pronk, J. L. Bos, and A. J. van der Eb. 1990. Analysis of the rat *JE* gene promoter identifies an AP-1 binding site essential for basal expression but not for TPA induction. *Nuc. Acids. Res.* 18: 23-34.
27. Yoshimura, T., N. Yuhki, S. K. Moore, E. Appella, M. I. Lerman, and E. J. Leonard. 1989. Human monocyte chemoattractant protein-1 (MCP-1) full-length cDNA cloning, expression in mitogen-stimulated blood mononuclear leukocytes, and sequence similarity to mouse competence gene *JE*. *FEBS Lett.* 244: 487-493.
28. Furutani, H. Nomura, M. Notake, Y. Oyamada, T. Fukui, M. Yamada, C. G. Larsen, J. J. Oppenheim, and K. Matsushima. 1989. Cloning and sequencing of the cDNA for human monocyte chemotactic an activating factor (MCAF). *Biochem. Biophys. Res. Comm.* 159: 249-255.
29. Decock, B., R. Conings, J.-P. Lenaerts, A. Billiau, and J. Van Damme. 1990. Identification of the monocyte chemotactic protein from human osteosarcoma cells and monocytes: detection of a novel N-terminally processed form. *Biochem. Biophys. Res. Comm.* 167: 904-909.
30. Chang, H. C., F. Hsu, G. J. Freeman, J. D. Griffin, and E. L. Reinherz. 1989. Cloning and expression of a gamma-interferon inducible gene in monocytes: a new member of a cytokine gene family. *Inter. Immunol.* 1: 388-397.
31. Graves, D. T, Y. L. Jiang, M. J. Williamson, and A. J. Valente. 1989. Identification of monocyte chemotactic activity produced by malignant cells. *Science* 245: 1490-1493.
32. Rollins, B. J., P. Stier, T. Ernst, and G. G. Wong. 1989. The human homolog of the *JE* gene encodes a monocyte secretory protein. *Mol. Cell. Biol.* 9: 4687-4695.
33. Yoshimura, T., E. A. Robinson, S. Tanaka, E. Appella, and E. J. Leonard. 1989. Purification and amino acid analysis of two human monocyte chemoattractant produced by phytohemagglutinin-stimulated human blood mononuclear leukocytes. *J. Immunol.* 142: 1956-1962.
34. Obaru, K., M. Fukuda, S. Maeda, and K. Shimada. 1986. A cDNA clone used to study mRNA inducible in human tonsillar lymphocytes by a tumor promoter. *J. Biochem.* 99: 885-894.

35. Nakao, M., H. Nomiyama, and K. Shimada. 1990. Structures of human genes coding for cytokine LD78 and their expression. *Mol. Cell. Biol.* **10**: 3646-3658.
36. Davatelis, G., P. Tekamp-Olson, S. D. Wolpe, K. Hermsen, C. Luedke, C. Gallegos, D. Coit, J. Merryweather, and A. Cerami. 1988. Cloning and characterization of a cDNA for murine macrophage inflammatory protein (MIP), a novel monokine with inflammatory and chemokinetic properties. *J. Exper. Med.* **167**: 1939-1944.
37. Sherry, B., P. Tekamp-Olson, C. Gallegos, D. Bauer, G. Davateliks, S. D. Wolpe, F. Masiarz, D. Coit, and A. Cerami. 1988. Resolution of the two components of macrophage inflammatory protein 1, and cloning and characterization of one of those components, macrophage inflammatory protein 1β. *J. Exper. Med.* **168**: 2251-2259.
38. Chang, H. C., and E. L. Reinherz. 1989. Isolation and characterization of a cDNA encoding a putative cytokine which is induced by stimulation via the CD2 structure on human T lymphocytes. *Eur. J. Immunol.* **19**: 1045-1051.
39. Lipes, M. A., M Napolitano, K.-T. Jeang, N. T. Chang, and W. J. Leonard. 1988. Identification, cloning and characterization of an immune activation gene. *Proc. Natl. Acad. Sci. USA* **85**: 9704-9708.
40. Brown, K. D., S. M. Zurawski, T. R. Mosmann and G. Zurawski. 1989. A family of small inducible proteins secreted by leukocytes are members of a new superfamily that includes leukocyte and fibroblast-derived inflammatory agents, growth factors and indicators of various activation processes. *J. Immunol.* **142**: 679-687.
41. Schall, T. J., J. Jongstra, B. J. Dyer, J. Jorgensen, C. Clayberger, M. M. Davis and A. M. Krensky. 1988. A human T cell-specific molecule is a member of a new gene family. *J. Immunol.* **141**: 1013-1025.
42. Burd, P. A., G. J. Freeman, S. D. Wilson, M. Berman, R. DeKruyff, P. R. Billings and M. E. Dorf. 1987. Cloning and characterization of a novel T cell activation gene. *J. Immunol.* **138**: 3126-3131.
43. Zipfel, P. F., J. Balke, S. G. Irving, K. Kelly and U. Siebenlist. 1989. Mitogenic activation of human T cells induces two closely related genes which share structural similarities with a new family of secreted factors. *J. Immunol.* **142**: 1582-1590.
44. Luster, A. D., J. C. Unkeless, and J. V. Ravetch. 1985. Gamma-interferon transcriptionally regulates an early-response gene containing homology to platelet proteins. *Nature* **315**: 672-676.
45. Luster, A. D. and J. V. Ravetch. 1987. Genomic characterization of a gamma-interferon inducible gene (IP-10) and identification of an interferon-inducible hypersensitive site, *Mol. Cell. Biol.* **7**: 3723-3731.
46. Matsushima, K., K. Morishita, T. Yoshimura, S. Lavu, Y. Kobayashi, W. Lew, E. Appella, H. F. Kung, E. J. Leonard, and J. J. Oppenheim. 1988. Molecular cloning of a human monocyte-derived neutrophil chemotactic factor (MDNCF) and the induction of MDNCF mRNA by interleukin 1 and tumor necrosis factor. *J. Exper. Med.* **167**: 1883-1893.
47. Mukaida, N., M. Shiroo, and K. Matsushima. 1989. Genomic structure of the human monocyte-derived neutrophil chemotactic factor IL-8. *J. Immunol.* **143**: 1366-1371.
48. Deuel, T. F., P. S. Keim, M. Farmer, and R. L. Heinrikson. 1977. Amino acid sequence of human platelet factor 4. *Proc. Natl. Acad. Sci. USA* **74**: 2256-2258.
49. Poncz, M., S. Surrey, P. LaRocco, M. J. Weiss, E. F. Rappaport, T. M. Conway, and E. Schwartz, Cloning and characterization of platelet factor 4 cDNA derived from a human erythroleukemic cell line. *Blood* **69**: 219-223.
50. Holt, J. C., M. E. Harris, A. M. Holt, E. Lange, A. Henschen, and S. Niewiarowski, Characterization of human platelet basic protein, a precursor form of low-affinity platelet factor 4 and β-thromboglobulin. *Biochemistry* **25**: 1988-1996.
51. Doi, T., S. M. Greenberg, R. D. Rosenberg, Structure of the rat platelet factor 4 gene: a marker for megakaryocyte differentiation. *Mol. Cell. Biol.* **7**: 898-904.
52. Green,C. J., R. St. Charles, B. F. P. Edwards, and P. H. Johnson. 1989. Identification and characterization of PF4var1, a human gene variant of platelet factor 4. *Mol. Cell. Biol.* **9**: 1445-1451.
53. Wenger, R. H., A. N. Wicki, A. Walz, N. Kieffer, and K. J. Clemetson. 1989. Cloning of cDNA coding for connective tissue activating peptide III from a human platelet-derived gt11 expression library *Blood* **73**: 1498-1503.
54. Begg, G. S., D. S. Pepper, C. N. Chesterman, and F. J. Morgan. 1978. Complete covalent structure of human β-thromboglobulin. *Biochemistry* **17**: 1739-1744.
55. Walz, A., and M. Baggiolini, A novel cleavage product of β-thromboglobulin formed in cultures of stimulated mononuclear cells activates human neutrophils. *Biochem. Biophys. Res. Comm.* **159**: 969-975.

56. Anisowicz, A., L. Bardwell, and R. Sager. 1987. Constitutive overexpression of a growth-regulated gene in transformed Chinese hamster and human cells. *Proc. Natl. Acad. Sci. USA* **84**: 7188-7192.
57. Richmond, A., E. Balentien, H. G. Thomas, G. Flaggs, D. E. Barton, J. Spiess, R. Bordoni, U. Francke, and R. Derynck. 1988. Molecular characterization and chromosomal mapping of melanoma growth stimulatory activity, a growth factor structurally related to β-thromboglobulin. *EMBO J.* 7:2025-2033.
58. Oquendo, P., J. Alberta, D. Wen, J. L. Graycar, R. Derynck, and C. D. Stiles. 1989. The platelet-derived growth factor-inducible *KC* gene encodes a secretory protein related to platelet α-granule proteins. *J. Biol. Chem.* **264**: 4133-4137.
59. Watanabe, K., K. Konishi, M. Fujioka, S. Kinoshita, and H. Nakagawa. 1989. The neutrophil chemoattractant produced by the rat kidney epithelioid cell line NRK-52E is a protein related to the *KC*/gro protein. *J. Biol. Chem.* **264**: 19559-19563.
60. Bedard, P.-A., D. Alcorta, D. L. Simmons, K.-C. Luk, and R. L. Erikson. 1987. Constitutive expression of a gene encoding a polypeptide homologous to biologically active human platelet protein in Rous sarcoma virus-transformed fibroblasts. *Proc. Natl. Acad. Sci. USA* **84**: 6715-6719.
61. Sugano, S., M. Y., Stoeckle, and H. Hanafusa. 1987. Transformation by Rous sarcoma virus induces a novel gene with homology to a mitogenic platelet protein. *Cell* **49**: 321-328.
62. Martins-Green, M., and M. J. Bissell. 1990. Localization of 9E3/CEF-4 in avian tissues: Expression is absent in rous sarcoma virus but is stimulated by injury. *J. Cell. Biol.* **110**: 581-595.
63. Wolpe, S. D., B. Sherry, D. Juers, G. Davatelis, R. W. Yurt and A. Cerami. 1989. Identification and characterization of macrophage inflammatory protein 2. *Proc. Natl. Acad. Sci. USA* **86**: 612-616.
64. Modi, W. S., M. Dean, H. N. Seuanez, N. Mukaida, K. Matsushima and S. J. O'Brien. 1990. Monocyte-derived neutrophil chemotactic factor (MDNCF/IL-8) resides in a gene cluster along with several other members of the platelet factor 4 gene superfamily. *Hum. Genet.* **84**: 185-187.
65. Wolpe, S. D. and A. Cerami. 1989. Macrophage inflammatory proteins 1 and 2: members of a novel superfamily of cytokines. *FASEB J.* 3: 2565-2573.

HUMAN NEUTROPHIL GRANULE CATIONIC PROTEIN CAP37 IS A SPECIFIC MACROPHAGE CHEMOTAXIN THAT SHARES HOMOLOGY WITH INFLAMMATORY PROTEINASES

John G. Morgan[1], H. Anne Pereira[2], Teresa Sukiennicki[1],
John K. Spitznagel[2], and James W. Larrick[1]

[1]Genelabs Inc.
505 Penobscot Drive, Redwood City, CA 94063
[2]Dept. of Microbiology and Immunology
Emory University School of Medicine, Atlanta GA 30322

ABSTRACT

Cationic antimicrobial protein CAP37 (Mr = 37 kD) is derived from the azurophilic granules of human PMN. In vitro and in vivo studies demonstrate that CAP37 is a novel monocyte-specific chemoattractant. The N-terminal amino acid sequence of CAP37 shares significant homology with a number of inflammatory molecules with protease activity including elastase and cathepsin G. However, substitutions in the catalytic triad (serine for a histidine at position 41 and glycine for a serine at position 175), may account for its lack of serine protease activity. A full length cDNA for CAP37 was identified in an HL60 cDNA library screened with oligonucleotide probes designed from the N-terminal amino acid sequence. Sequencing of the cDNA reveals a protein of 225 amino acids with significant nucleotide homology to cathepsin G and human neutrophil elastase.

INTRODUCTION

Inflammation is orchestrated by the sequential release of specific molecules that mediate the activation and chemotaxis of various cell types (1). The first wave of cells in early inflammatory lesions is comprised largely of polymorphonuclear leukocytes (PMN) followed by infiltration of mononuclear phagocytes. It has been hypothesized that one or more factors derived from the PMN may modulate the subsequent chemotaxis of monocytes (2,3). In fact an early study described the absolute requirement of PMNs for the subsequent infiltration of mononuclear phagocytes (4).

This laboratory demonstrated the antimicrobial activity of a major PMN granule protein CAP 37 [designated Cationic Antimicrobial Protein of Mr = 37kD](5,6). This protein is a potent mediator of monocyte specific chemotaxis, and thus may act as a mediator of the second wave of inflammation in vivo (7).

Recently several macrophage/monocyte chemotaxins have been described from other sources (for review see [8]). These have included: GDCF-2 (Glioma-derived chemotactic factor), MCAF (Monocyte chemotactic and activation factor),

Chemotactic Cytokines, Edited by J. Westwick *et al.*
Plenum Press, New York, 1991

LDCF (Lymphocyte-derived chemotactic factor). Collectively, these are now known as macrophage chemotactic peptide 1 (MCP1). This monocyte specific chemotaxin is active at nanomolar concentrations. It is a 12.5 kD protein composed of 76 amino acids with a 23 amino acid signal peptide. MCP1 is a member of the C-X-C...C...C gene family that is described in other chapters in this volume. While members of this family of inflammatory and tissue repair proteins share 25 to 55% homology at the amino acid level among themselves, they do not share homology with CAP37 and related proteins. They are widely expressed in various tissues and cells whereas CAP37 has a more restricted distribution. A schematic of our understanding of the relationship of these chemotactic proteins to first and second wave inflammation is presented in figure 1.

Immunocytochemical studies performed on peripheral blood cells indicate that CAP37 is present only in PMN. Mononuclear cells, eosinophils, and red blood cells do not stain positive with monospecific anti-CAP37 antibody. This same antiserum was used to demonstrate the presence of immunoreactive CAP37 in HL60 cells that were used for the cDNA studies described below.

This chapter summarizes various cellular and molecular properties of CAP37. This protein is related to the serine class of proteinases and shares significant homology with several other proteases derived from the granules of inflammatory leukocytes that are thought to mediate tissue damage during the course of inflammatory diseases (9).

MATERIALS AND METHODS

Details of all of the methods used to derived the data described below can be found in papers written by our groups in the past year (7,9).

RESULTS AND DISCUSSION

1. *CAP37 is a macrophage specific chemotaxin*

In vitro studies using a modified Boyden chamber demonstrated that purified CAP37 (but not elastase) was specifically chemotactic for human monocytes (see figure 2). Significant chemotaxis was apparent at 100 pM concentration. Details of these studies have been presented (7).

CAP37 is a potent chemotaxin in vivo. When purified CAP37 was injected IP into mice a significant emigration of monocytes was found versus saline control (see figure 3).

2. *CAP37 is a member of the serine proteinase family*

The N-terminal sequence of CAP37 shares significant homology with a number of serine proteinases (see figure 4). The first 20 residues of CAP37 were similar to the first 20 amino acid residues of the recently published sequence of an antibacterial substance named azurocidin (10). Azurocidin was purified from the membranes of azurophil granules of human PMN reported to have a molecular weight of 29 kD. Despite the disparity in the molecular weights, CAP37 and azurocidin appear similar.

The serine endopeptidases are the largest class of mammalian proteinases. The sequence of the CAP37 protein expands the number of serine endopeptidase-homologues associated with inflammatory cells. The elastase/medullasin cDNA encodes a protein of 237 amino acid residues (11). CAP37 shares highest protein and

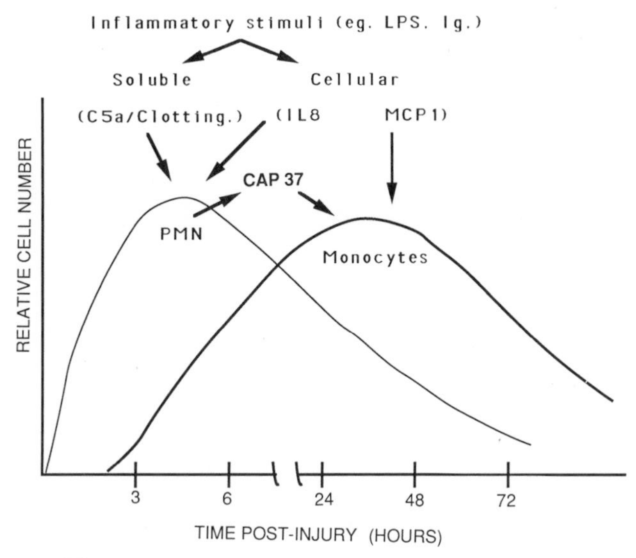

Figure 1. Kinetics of inflammatory cell chemotaxis.

and nucleic acid homology with this protein (12,13,14). PMN elastase is a proteolytic enzyme with broad substrate specificity that resides in the azurophil granules of PMNs and monocytes. It is a 28 kd protein that removes the hyaluronic acid-binding region of cartilage proteoglycan and then fragments the glycosaminoglycan attachment region (15). Although elastin is generally resistant to proteolytic attack, it is degraded by elastase (16). PMN elastase also degrades fibronectin and laminin (17), and type I, type II and type IV collagens (18). The activity of elastase is modulated in vivo by alpha-1-proteinase inhibitor and alpha-2-macroglobulin. Two clinically important anti-inflammatory drugs, gold sodium thiomalate and pentosan polysulfate, act in part by inhibiting leukocyte elastase (19).

Cathepsin G, also present in the azurophil granules, is structurally and catalytically related to the pancreatic chymotrypsins and to the chymases of mast cell granules. Although cathepsin G has limited action on elastin or type I collagen, it can solubilize collagen from cartilage and may generate physiologically active components from the complement proteins (20). Plasma alpha-1 anti-chymotrypsin and alpha-1-proteinase inhibitor and alpha-2-macroglobulin are the physiological inhibitors of cathepsin G (21,22).

As shown in figure 4 the cys-26 residue of CAP37 which corresponds to the cys-42 residue of the first disulfide bond (cys-42 to cys-58 in chymotrypsin numbering) of serine proteases is conserved. The cDNA demonstrates that the other cysteines are also conserved between CAP37, cathepsin G and neutrophil elastase.

3. CAP37 is an inactive proteinase

In addition to the conserved disulfide bonds, another important feature of the serine proteases is the residues forming the 'charge relay system' of the active catalytic site which occurs at his-57, asp-102 and ser-195 (chymotrypsin numbering). The protein sequence and cDNA sequence of CAP37 demonstrate that the conserved histidine (#41) is replaced by a serine residue, the aspartic acid (#89) is conserved and the serine (#175) is replaced by a glycine (see figure 5).

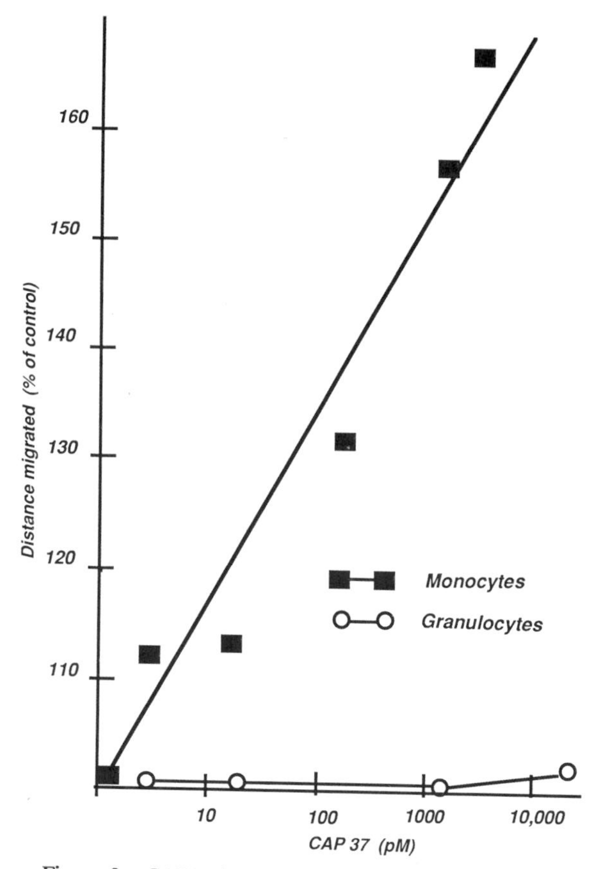

Figure 2. CAP37 is a monocyte specific chemotaxin

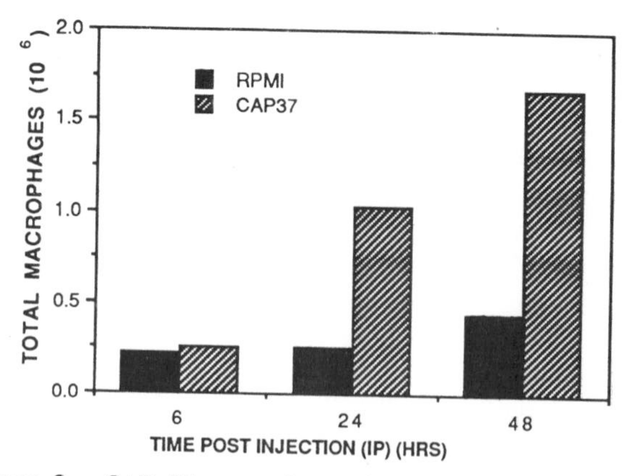

Figure 3. CAP 37 macrophage chemotaxis in vivo: specific chemotaxis of monocytes into peritoneal cavity of mice injected with purified CAP37 (1 μg)

```
               1                   10                  20
CAP37          I-V-G-G-R-K-A-R-P-R-Q-F-P-F-L-A-S-I-Q-N-Q-G-R-H-F---
ELAST          I-V-G-G-R-R-A-R-P-H-A-W-P-F-M-V-S-L-Q-L-R-G-G-H-F---
FACTD          I-L-G-G-R-E-A-E-A-H-A-R-P-Y-M-A-S-V-Q-L-N-G-A-E-L---
PLASM          I-V-G-G-C-V-S-K-P-H-S-W-P-W-Q-V-S-L-R-R-S-S-R-H-F---
CATG           I-I-G-G-R-E-S-R-P-H-S-R-P-Y-M-A-Y-L-Q-I-Q-S-P-A-G-Q-
RMCPI          I-I-G-G-V-E-S-R-P-H-S-R-P-Y-M-A-H-L-E-I-T-T-E-R-G-Y-
RMCPII         I-I-G-G-V-E-S-I-P-H-S-R-P-Y-M-A-H-L-D-I-V-T-E-K-G-L-
CCPI           I-I-G-G-H-E-V-K-P-H-S-R-P-Y-M-A-L-L-S-I-K-D-Q-Q-P-E-
HF             I-I-G-G-D-T-V-V-P-H-S-R-P-Y-M-A-L-L-K-L-S-S-N-T------

                               30                  40
                          *                      + *
CAP37                   --C-G-G-A-L-I-H-A-R-F-V-M-T-A-A-S---  41
ELAST                   --C-G-A-T-L-I-A-P-N-F-V-M-S-A-A-H-C-  42
FACTD                   --C-G-G-V-L-V-A-E-Q-W-V-L-T-A-A-H-C-  42
PLASM                   --C-G-G-T-L-I-S-P-K-W-V-L-T-A-A-H-C-  42
CATG               -S-R---C-G-G-F-L-V-R-E-D-F-V-L-T-A-A-H-C-  45
RMCP1              -K-A-T-C-G-G-F-L-V-N-R-N-F-V-M-T-A-A-H-C-  46
RMCPII             -R-V-I-C-G-G-F-L-I-S-R-Q-F-V-L-T-A-A-H-C-  46
CCPI               -A---I-C-G-G-F-L-I-R-E-D-F-V-L-T-A-A-H-C-  45
HF                      I-C-A-G-A-L-I-E-K-N-W-V-L-T-A-A-H-C-  42
```

Figure 4. Alignment of the amino acid sequences of CAP37 with several inflammatory cell proteases.
ELAST (elastase), FACTD (complement factor, D), PLASM (bovine plasminogen), CATG (cathepsin G), RMCPI (rat mast cell, protease I), RMCP II (rat mast cell protease II), CCPI (cytotoxic T cell protease I), HF (cytotoxic T cell protease, H factor); * indicates position of disulfide bonds. + indicates expected location of 'his' residue of the serine protease catalytic triad.

Because of the strong homologies between CAP37 and the other serine proteases it was important to determine if CAP37 would bind the serine protease inhibitor DFP, and whether it was capable of degrading various substrates specific for serine proteases. Figure 6 summarizes previously published work showing that CAP37 does not bind to tritiated DFP and does not cleave a number of substrates of elastase and cathepsin G.

4. cDNA clone of CAP37

Figure 7 presents a schematic diagram of the CAP37 cDNA clone. This clone was identified in an HL60 cDNA library using olionucleotide probes designed from the N-terminal sequence of the protein. Details of this work can be found elsewhere [Morgan et al., submitted to J. Biol. Chem.]. The salient features of this clone are:

	Enzymatic Activity	Substitutions
ELASTASE:	Active	His^{41} --- Asp^{88} --- Ser^{173}
CAP37:	Inactive	Ser^{41} --- Asp^{89} --- Gly^{175}

Figure 5: Mutations in the serine protease catalytic triad

ASSAY*	CAP37	ELASTASE	CATHEPSIN G
^{3}H-DFP binding	--	+++	+
Azocasein Degradation	--	++	+++
Hemoglobin Degradation	--	+++	++
BLT esterase Activity	--	+++	+++
Cleavage of Suc-Ala$_3$-Pro-Phe-pNA Suc-Ala$_{3\text{-}pNA}$	--	+++	--
Monocyte Chemotaxis	+++	--	--

*See Pereira et al. (1990) for details.

Figure 6: CAP37 is an inactive serine protease

4.1. The leader sequence contains three potential signal peptide cleavage sites. This suggests that CAP37 might be synthesized as a preproprotein. Further investigation will be required to test this idea.

4.2. Mutations are present is the consensus catalytic triad as noted above (figure 5). This finding corroborates biochemical data showing that CAP37 does not have proteolytic activity. Future investigations will combine molecular modeling and site-specific mutagenesis to modify these and other residues to determine if it is possible to restore the catalytic activity of the molecule.

4.3. The CAP37 cDNA encodes a protein of 225 amino acids with three potential glycosylation sites and 8 cysteines. The calculated molecular weight of the unmodified protein is 24,276 daltons and the pI is consistent with the known cationic charge of the molecule.

4.4. The CAP37 message of 904 bp is expressed only in immature myeloid cells (data not shown).

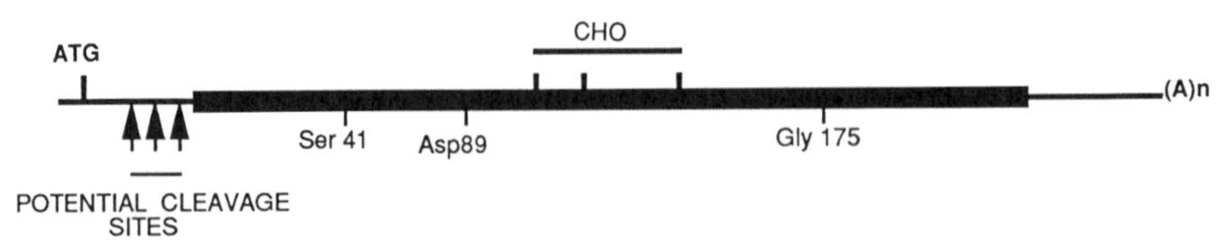

Figure 7: Schematic diagram of the CAP37 cDNA showing the positions of the catalytic triad, signal cleavage sites and N-linked carbohydrate addition sites.

CONCLUSIONS

Studies of the cloned cDNA for CAP37 will help to elucidate the role of this molecule in the inflammatory response. The demonstrated monocyte chemotactic properties of CAP37 suggest that it may promote wound healing by attracting these cells to areas of tissue damage. Expression of this molecule combined with purification of substantial quantities of recombinant protein will be required to determine the substrate specificity (if any) of the CAP37 molecule. This will also permit study of its regulation during inflammation and the relevance of its binding to endotoxin (27) to its antimicrobial action. It is likely that CAP37 will be differentially regulated during myeloid differentiation like other azurophilic granule proteins.

Acknowledgements: This work was supported by Public Health Service grant AI17662 (JKS) from the National Institute of Allergy and Infectious Disease.

REFERENCES

1. Wilkinson, P.C. 1982. Locomotion and chemotaxis of individual leucocyte types. *Chemotaxis and inflammation,* 2nd Ed. Edinburgh: Churchill Livingstone, 119-135.
2. Ward, P.A. 1968. Chemotaxis of mononuclear cells. *J. Exp. Med.* 128: 1201-1210.
3. Gallin, J.I. 1985. Neutrophil specific granule deficiency. *Ann. Rev. Med.* 36: 263-282.
4. Page, A.R., R.A. Good. 1958. A clinical and experimental study of the function of neutrophils in the inflammatory response. *Amer. J. Pathol.* 34: 645-655.
5. Shafer, W.M., L.E. Martin, J.K. Spitznagel. 1986. Late intraphagosomal hydrogen ion concentration favours the in vitro antimicrobial capacity of a 37-kilodalton cationic granule protein of human neutrophil granules. *Infect. Immun.* 53: 651-655.
6. Modrzakowski, M.C., J.K. Spitznagel. 1979. Bactericidal activity of fractionated granule contents from human polymorphonuclear leucocytes: Antagonism of granule cationic proteins by lipopolysaccharide. *Infect. Immun.* 25: 597-602.
7. Pereira, H.C., W.M. Shafer, J. Pohl, L.E. Martin, J.K. Spitznagel. 1990. CAP37, a human neutrophil-derived chemotactic factor with monocyte specific activity. *J. Clin. Invest.* 85: 1468-1476.
8. Leonard, E.J., T. Yoshimura. 1990. Human monocyte chemoattractant protein-1 (MCP-1). *Immunology Today.* 11: 97-100.
9. Pereira, H.C., J.K. Spitznagel, J. Pohl, D. Wilson, W.M. Shafer, J. Morgan, I. Palings, J.W. Larrick. 1990. The 37 KD human neutrophil cationic protein shares homology with inflammatory proteinases. *Life Sciences.* 46: 189-196.
10. Gabay, J.E., R.W. Scott, D. Campanelli, J. Griffith, C. Wilde, M.N. Marra, M. Seeger, C.F. Nathan. 1989. Antibiotic proteins of human polymorphonuclear leukocytes. *Proc. Natl. Acad. Sci. USA.* 86: 5610-5614.
11. Salvesen, G., D. Farley, J. Shuman, A. Przybyla, C. Reily, J. Travis. 1987. Molecular cloning of human cathepsin G. Structural similarity to mast cell and cytotoxic lymphocyte proteinases. *Bio. Chem.* 26: 2289-2295.
12. Sinha, W., W. Watorek, S. Karr, J. Giles, W. Bode, J. Travis. 1987. Primary structure of human neutrophil elastase. *Proc. Natl. Acad. Sci.* 84: 2228-2223.
13. Okano, K., Y. Aoki, Sakurai, M. Kajitani, S. Kanai, T. Shimazu, H. Shimizu, M. Nruto. 1987. Molecular cloning of complementary DNA for human medullaisn; an inflammatory serine protease in bone marrow cells. *J. Biol. Chem.* 102: 13-19.
14. Farley, D., G. Salvensen, J. Travis. 1988. Molecular cloning of human neutrophil elastase. *Bio. Chem. Hoppe. Seyler.* 369: 3-11. (suppl.).
15. Keiser, H., R.A. Greenwald, G. Feinstein, A. Janoff. 1976. Degradation of cartilage proteoglycan by human leukocyte neutral proteases. A model of joint injury. II. Degradation of isolated bovine nasal cartilage proteoglycan. *J.Clin.Invest.* 57: 625-632.
16. Barrett, A.J. 1981. Leukocyte elastase. *Methods Enzymol.* 80: 581-610.
17. Cambell, E.J., R.M. Senior, J.A. McDonald, D.L. Cox. 1982. Proteolysis by neutrophils. Relative importance of cell-substrate contact and oxidative inactivation of proteinase inhibitors in vitro. *J. Clin. Invest.* 70: 845-852.

18. Pipoly, D.J., E.C. Crouch. 1987. Degradation of native type IV procollagen by human neutrophil elastase. Implications for leucocyte-mediated degradation of basement membranes. *Biochem.* **26**: 5748–5755.
19. Baici, A., P. Salgam, K. Fehr, A. Boni. 1981. Inhibition of human elastase from polymorphonuclear leucocytes by gold sodium thiomalate and pentosan polysulfate. *Biochem. Pharmacol.* **30**: 703–710.
20. Starkey, P.M., A.J. Barrett, M.C. Burleigh. 1977. The degradation of articular collagen by neutrophil proteinases. *Biochim. Biophys. Acta.* **483**: 386–392.
21. Ohlsson, K., L. Ohlsson. 1974. Neutral proteases of human granulocytes. III. Interaction between granulocyte elastase and plasma protease inhibitors. *Scand. J. Clin. Lab. Invest.* **34**: 349–357.
22. Barrett, A.J., P.M. Starkey. 1973. The interaction of alpha-2 macroglobulin with proteinases. Characteristics and specificity of the reaction and a hypothesis concerning its molecular mechanism. *Biochem. J.* **133**: 709–715.

BIOCHEMICAL AND BIOLOGICAL CHARACTERIZATION OF NAP-1/ IL-8-RELATED CYTOKINES IN LESIONAL PSORIATIC SCALE

Jens-M. Schröder

Department of Dermatology
University of Kiel
2300 Kiel, West Germany

INTRODUCTION

Psoriasis is a skin disease characterized by epidermal hyperproliferation as well as an inflammatory cell infiltrate in the epidermis predominantly consisting of polymorphonuclear leukocytes (PMN) (1). The absence of bacteria as the cause of PMN infiltration led to the hypothesis that endogenous chemoattractants located in the epidermal cell layer of psoriatic plaques may attract PMN into the inflammatory focus.

Initial analysis of PMN-chemotactic factors in psoriatic skin revealed proteinaceous attractants like complement-derived chemotaxins (2), as well as lipid-like chemotaxins like LTB_4-related material (3), 12-HETE (4) as well as PAF (5).

These findings caused us to further characterize the relative amounts of the non-lipid-like PMN-chemoattractant material. Initial experiments had shown that trypsin treatment destroyed the majority of PMN-chemotactic activity indicating that PMN chemoattractant(s) are proteins. When an aqueous psoriatic scale extract was heated to 70°C PMN chemotactic activity was almost unaffected. These findings led to the conclusion that chemotactic protein(s) present in psoriatic scales are of relative heat resistance.

PROTEINACEOUS NEUTROPHIL ATTRACTANTS IN PSORIASIS

When we investigated extracts of lesional psoriatic scales, constantly PMN chemotactic activity was detected in ether, as well as aqueous, extracts. Interestingly, however, we observed that the majority of the activity apparently was present in the aqueous extracts, i.e. whereas 10 µl of the reconstituted lipid extract were effective to stimulate PMN chemotaxis half maximally, only 0.3 µl of the aqueous (lipid depleted) psoriatic scale extract were necessary to elicit half maximal PMN chemotaxis.

The only PMN chemotaxin previously known to be heat resistant and of proteinaceous nature had been identified as complement split product C5a (6). Therefore it appeared to be likely to assume that chemoattractant material obtained from psoriatic scale extracts is identical with a split product of the 5th complement component. Indeed, previous investigations of the "psoriatic leukotactic factor, PLF" revealed that PLF contains a fragment of C5 (7,8).

Chemotactic Cytokines, Edited by J. Westwick *et al.*
Plenum Press, New York, 1991

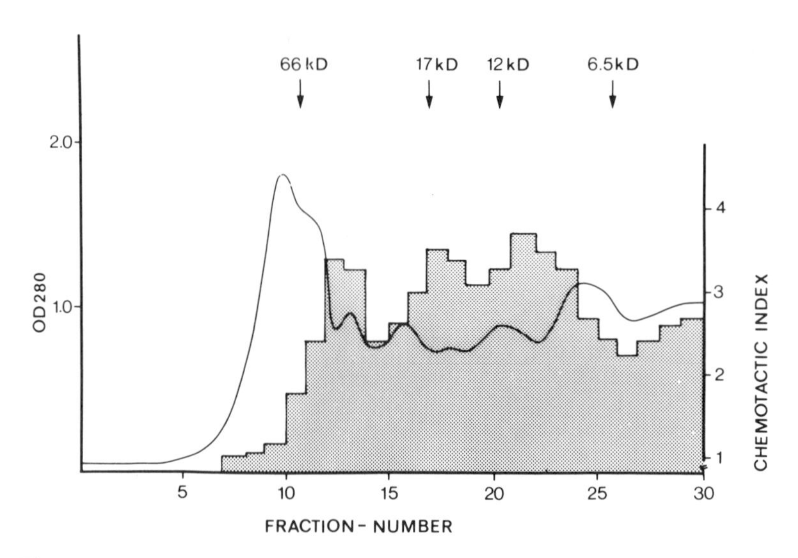

Figure 1. G-75 gel chromatography of psoriatic scale extract. An acidic extract of psoriatic scales was chromatographed on a G-75 gel column and the effluent analyzed for UV absorption (OD_{270}, -) or PMN chemotactic activity (⁙). Elution positions of proteins with known M_r are indicated by arrows.

The final conclusion however, that PLF is identical with the anaphylatoxin C5a (7) appeared to be questionable because the anaphylatoxin C5a is known to be inactivated *in vivo* within seconds by carboxypeptidase N (9). Therefore we re-investigated the nature of proteinaceous chemoattractants obtained from lesional psoriatic scales.

When extracts obtained from pooled psoriatic scales were applied to a G-75 gel column, constantly several peaks of PMN chemotactic activity were obtained (Figure 1). Whereas peak-PMN chemotactic activity eluting in fractions no. 17-19 could be characterized as $C5a_{des\ arg}$ (8), activity eluting in fractions no. 20-25 from the column appeared to derive from a hitherto in psoriatic lesions not discovered PMN chemotaxin.

DETECTION OF A NOVEL PMN CHEMOTACTIC FACTOR

In order to purify this apparently novel PMN chemotactic protein we first investigated binding characteristics to an ion exchanger. It was shown that, unlike $C5a_{des\ arg}$, this activity apparently does not bind to a cation exchanger at pH7.8. Because of its more anionic behavior (relative to $C5a_{des\ arg}$) we tentatively termed this neutrophil chemotactic factor "anionic neutrophil activating peptide, ANAP" (8).

G-75 gel chromatography analyses of psoriatic scale extracts obtained from single patients revealed that ANAP apparently is constantly present in all extracts investigated, whereas sometimes a decreased amount – in some extracts also an absence – of $C5a_{des\ arg}$ was observed (data not shown). Therefore we tried to further characterize ANAP.

When partially purified (by cation exchange), $C5a_{des\ arg}$-depleted ANAP was analyzed for cross desensitization of chemotaxin receptor-dependent enzyme release, no cross-reactivity with well characterized PMN chemotactic factors like C5a, LTB_4, FMLP or PAF was observed (Table 1). These findings indicated that ANAP apparently binds to a PMN receptor, which is distinct from that described for

Table 1. Cross-desensitization of PMN degranulation elicited with ANAP and other PMN Chemotactic Factors

Preincubation with	Stimulation with ANAP	FMLP 10^{-8}M	C5a 10^{-8}M	LTB_4 10^{-8}M	PAF 10^{-6}M	NAP-1/ IL-8 10^{-8}M
ANAP	93.3	-3.3	3.4	0.3	-6.2	86.4
BocMLP 10^{-5}M	4.6	89.5	0.6	-2.4	4.2	1.4
C5a $2x10^{-8}$M	1.3	7.2	85.3	2.7	1.4	-3.4
LTB_4 $2x10^{-8}$M	-3.7	2.2	0.3	89.6	3.1	-4.7
PAF $2x10^{-6}$	18.6	-23.4	-11.2	-26.8	96.7	-21.1
NAP-1/IL-8 $2x10^{-8}$M	96.1	4.4	2.8	-3.1	0.2	96.1

Enzyme release (β-glucuronidase) was determined in chemotaxin pre-incubated PMN after subsequent stimulation with different chemotaxins. Results are expressed in percent inhibition of control enzyme release (PMN treated with PBS).

FMLP, C5a, LTB_4 or PAF. Interestingly, PMN pre-incubated with ANAP did not release β-glucuronidase when restimulated with a PMN chemotactic protein produced by mononuclear cells, when stimulated with LPS (Table I).

This factor was recently purified to homogeneity in different laboratories (10-13) and is now termed NAP-1/IL-8 (14,15). PMN receptor cross-reactivity of ANAP with NAP-1/IL-8 led to the hypothesis, that ANAP might be biochemically similar or identical with NAP-1/IL-8.

The earlier conclusion that neutrophil attractant material obtained from psoriatic scales was identical with IL-1 or the epidermal cell thymocyte activating factor ETAF (16,17), both shown to be chemotactically active in a partially purified form (16,17) is erroneous, because ANAP activity and IL-1/ETAF activity could be separated from each other (Figure 2). Moreover, recombinant human IL-1α and IL-1β have been shown to lack neutrophil chemotactic and activating properties (18), which supports the suggestion that ANAP is a different cytokine.

PURIFICATION OF ANAP

Although initial efforts to purify ANAP to homogeneity were not successful because of the lack of psoriatic scale material as well as high amounts of contaminating proteins, we developed an extraction procedure which allowed the use of HPLC as the first step to purify ANAP. Routinely 50-100 grams of pooled lesional scale material were extracted at acidic pH (pH3) with aqueous ethanol. Such conditions have been proved to be optimal in our laboratory. The major advantage of this procedure has been found to be the separation of large amounts

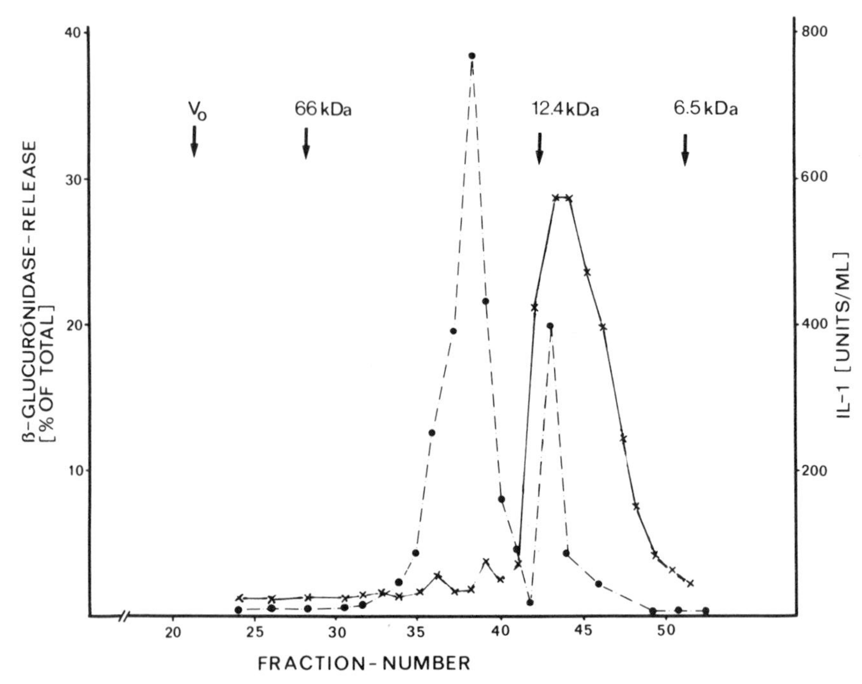

Figure 2. Dissociation of neutrophil activating factors and interleukin 1 by TSK-2000-HPLC. Partially purified ANAP was separated by size exclusion HPLC and fractions tested for enzyme releasing activity (—) as well as IL-1 activity (- -). Note the absence of co-elution of both activities.

of high molecular weight proteins allowing a cation exchange HPLC as the first step to purify ANAP (Figure 3).

To our surprise two broad peaks of PMN chemotactic activity could be eluted from the HPLC column when increased amounts of ammonium formate were used to elute proteins (Figure 3). Lowering the pH eluted the majority of chemotactic material (Figure 3). The broad peak of PMN chemotactic activity eluting first from the column tentatively was termed α-ANAP, whereas the chemotactic polypeptides eluting at lower pH and higher ion strength were termed β-ANAP (Figure 3).

STRUCTURAL CHARACTERIZATION OF THE MAJOR ANAPs

We were able to purify nine different ANAP factors by the use of various HPLC techniques including preparative wide pore reversed phase (RP-8) HPLC, cyanopropyl HPLC, poly-F HPLC, TSK 2000 size exclusion HPLC as well as analytical narrow pore RP-18 HPLC (19, and data not shown). All ANAP peptides were found to be of apparent homogeneity shown by the presence of a single line upon SDS PAGE (Figure 4 and data not shown).

The quantitatively dominating ANAP factors, β_1-ANAP, α_2-ANAP and β_2-ANAP, could be obtained in sufficient amounts for aminoterminal amino acid sequence analysis. As a result we found that the quantitatively by far dominating ANAP was β_1-ANAP, which shows an aminoterminal sequence identical to that found for the 69 residue form of NAP-1/IL-8, Table 2 (19).

α_2-ANAP has been found to contain an aminoterminal sequence identical with that published for the "16 kD moiety of melanoma growth stimulatory activity

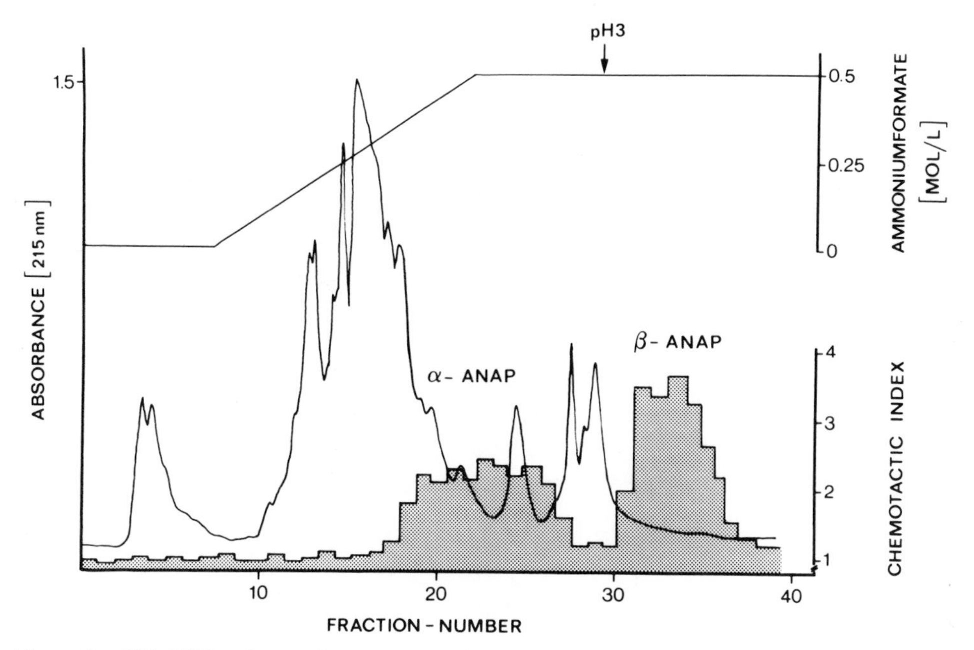

Figure 3. CM-TSK cation exchange HPLC of a psoriatic scale extract. An acidic and ethanolic extract of psoriatic scales was separated by cation exchange HPLC. Proteins were eluted using a gradient of increasing concentration of ammonium formate and finally decrease of the pH. Fractions were tested at 100-fold dilution for PMN chemotactic activity using the indirect cell counting method recently described (10). Note the presence of two broad peaks of biological activity. The peak eluting first is termed α-ANAP, whereas the second peak contains β-ANAP.

MGSA/gro", Table 2 (19,20). β_2-ANAP showed the same aminoterminal amino acid sequence as found for the 72 residue form of NAP-1/IL-8 as well as the same mobility upon SDS PAGE (Figure 4) and is therefore identical with NAP-1/IL-8, Table 2 (19).

When PMN chemotaxis of purified β_1-ANAP, β_2-ANAP and α_2-ANAP was investigated, a dose-dependent response was observed (Figure 5). Interestingly all three ANAP factors stimulated PMN chemotaxis with similar ED_{50}, which have been found to be identical with the ED_{50} calculated for authentic NAP-1/IL-8 (Figure 5). Moreover, all major ANAP's were able to elicit the release of lactoferrin in PMN (Figure 6) and therefore should be considered as one type of the cytokines responsible for enhanced lactoferrin levels in psoriasis (21). These

Table 2. Aminoterminal amino acid sequence of some ANAPs*

β_1-ANAP		ELRXQXIKTYSKPFHPK...
β_2-ANAP		SAKELRXQXIKTYSKPFHPK...
β_2-ANAP	47%	SAKELRXQXIKTYSKPFHPK...
	43%	AVLPRSAKELRXQXIKTYSKPFHPK...
NAP-1/IL-8		SAKELRCQCIKTYSKPFHPKFIKELRVIESGPHCA...
α_2-ANAP		XXVATELRXQXLQTLQG...
MGSA/gro		ASVATELRXQXLQTLQG-IHPKNIQSVMVKSPGPHCA...

* the single letter code of amino acids was used.
X: amino acid could not be determined.

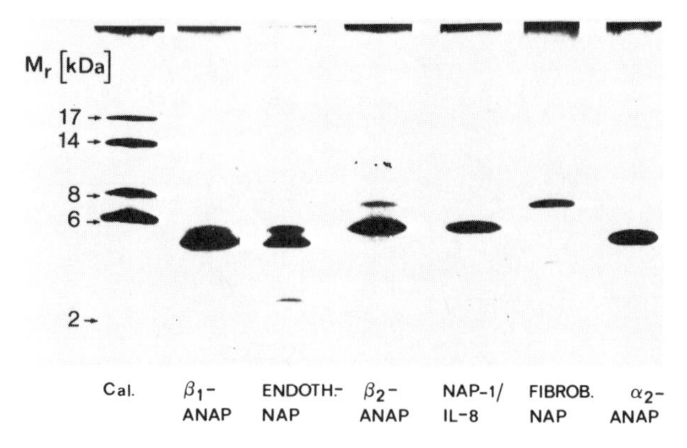

Figure 4. SDS PAGE analysis of some purified ANAP preparations. Purified ANAP preparations as well as some authentic neutrophil-activating proteins with known amino acid sequence were separated by SDS PAGE for small proteins in the presence of urea using the system of Schägger and von Jagow (32). Note identity of β_1-ANAP with an endothelial cell-derived NAP (22) and identity of β_2-ANAP with authentic NAP-1/IL-8. β_2-ANAP contains traces of the 77 residue form of NAP-1/IL-8 known to be produced by fibroblasts (23). α_2-ANAP shows a mobility similar to that seen for β_1-ANAP.

findings indicate that NAP-1/IL-8 and related attractants are present in psoriatic lesions in a biologically active form. As mentioned above, the dominating proteinaceous PMN attractant in psoriatic scales is the 69 residue form of NAP-1/IL-8 (β_1-ANAP). We have isolated the same NAP-1/IL-8-related polypeptide from LPS-stimulated endothelial cells (22,23) as well as from supernatants of growing keratinocytes (unpublished results).
So far it is not known which cell type is the major producer of β_1-ANAP. Although it appears to be likely that keratinocytes are the major source, in our hands cultivated keratinocytes were only poor producers of NAP-1/IL-8-related attractants even when stimulated with IL-1 or TNFα (unpublished results).

Recent studies, however, have shown that under certain circumstances, i. e. pretreatment of keratinocytes with γ-interferon and subsequent stimulation with TNFα (24) or stimulation with urushiol, the skin-toxic principle of toxic ivy (25), high amounts of the NAP-1/IL-8 gene were expressed. On the other hand it is also possible that the 77 residue form of NAP-1/IL-8, known to be produced in large amounts by IL-1α-stimulated dermal fibroblasts (23), will be converted by keratinocyte proteases. Some of our β_2-ANAP preparations indeed contained in varying amounts the 77 residue form of NAP-1/IL-8 (Table 2).

ONE ANAP-FACTOR IS IDENTICAL WITH MGSA/GRO

α_2-ANAP (MGSA/gro), which is present in similar amounts as found for β_1-ANAP in psoriatic scales, belongs to the same family of β-thromboglobulin-like host defense cytokines (26,27). Although in PMN chemotaxis the ED_{50} of α_2-ANAP is similar to that found for NAP-1/IL-8, the efficiency (number of migrating PMN) at optimal concentration is lower (Figure 5). Moreover at high doses PMN did not respond to α_2-ANAP.

Similar findings we observed with a recently characterized neutrophil attractant initially termed NAP-3, which is identical to melanoma growth stimulatory activity

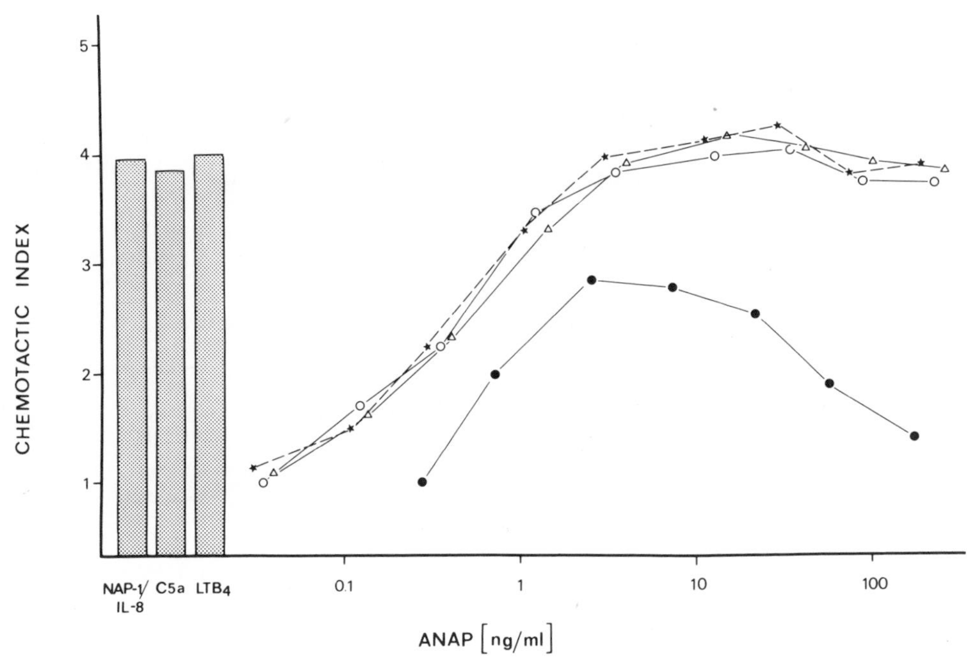

Figure 5. PMN chemotactic activity of ANAP preparations. β_1-ANAP (○), β_2-ANAP (✫) and α_2-ANAP (●) were tested for dose-dependent chemotactic activity using the indirect Boyden chamber assay. For comparison, Boyden chamber chemotaxis of authentic NAP-1/IL-8 is presented also (△). Note the similarity of dose response curves for β_1-ANAP, β_2-ANAP and NAP-1/IL-8 and the similar ED_{50}, however lower efficiency (number of migrating PMN) of α_2-ANAP.

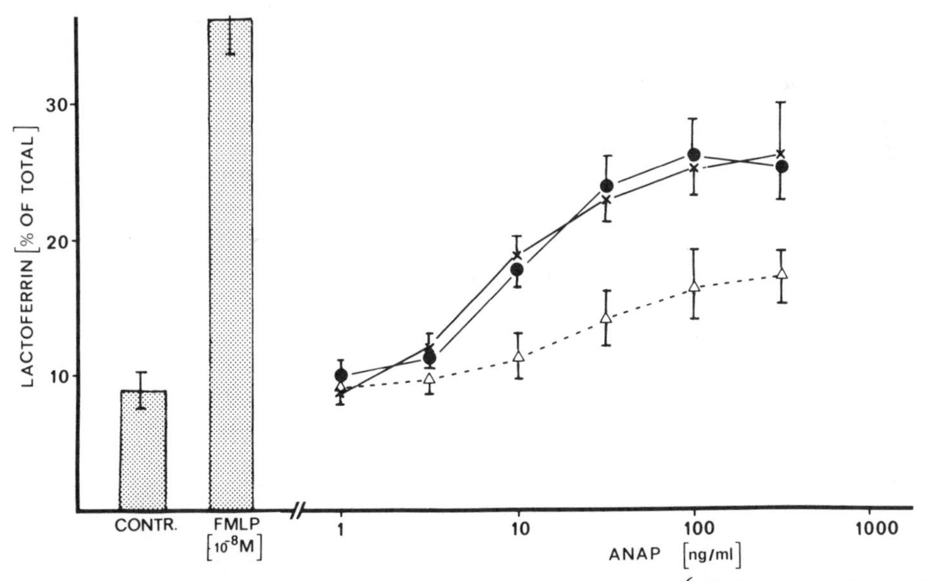

Figure 6. Release of lactoferrin by PMN stimulated with ANAP. 5 x 10^6 PMN were incubated with β_1-ANAP (x), β_2-ANAP (●) and α_2-ANAP (△), for 30min, respectively. Supernatants were analyzed for lactoferrin by the use of an ELISA (21). Results are expressed as the result of a total control (threefold freeze thawing). Note the presence of an ANAP dose-dependent lactoferrin release by PMN.

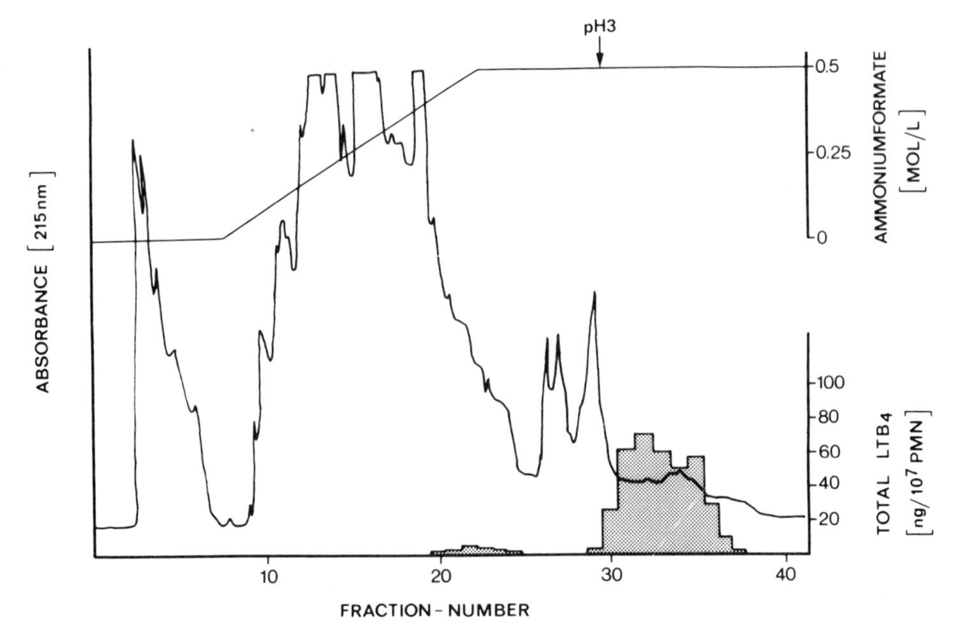

Figure 7. Detection of LTB_4-synthesis inducing activity in PMN by psoriatic scale products separated by cation exchange HPLC. Psoriatic scale extracts obtained by extraction with aqueous ethanol at pH3 were applied to a cation exchange HPLC column (CM TSK 3SW) and proteins were eluted using a gradient of increasing concentration of ammonium formate (pH5) as well as by lowering the pH. Fractions were lyophilized and separately incubated for 10 min. with 10^7 PMN in the presence of 10^{-5}M exogenous arachidonic acid. Supernatants were analyzed by RP-HPLC and the total amount of LTB_4 produced (sum of LTB_4 + ω-OH-LTB_4 + ω-COOH-LTB_4) was calculated (⁚⁚⁚). In the absence of exogenous arachidonic acid no significant LTB_4 was observed (not shown).

MGSA/gro (28). The only difference between α_2-ANAP and NAP-3 is the presence of two unidentifiable residues in the amino terminus of α_2-ANAP. Therefore these both closely related MGSA/gro forms appear to be biologically equipotent. Reasons for the failure to identify the amino acids in the amino terminus of α_2-ANAP, are so far unknown. It is known from MGSA, that the second amino acid is a Ser residue. Therefore O-derivatisation could explain this behavior. Moreover when analyzing PMN attracting polypeptides from LPS-stimulated monocytes we could identify a MGSA/gro form, which contains two serines in its both amino terminal residues. Possibly a derivatized form of such MGSA/gro variant may account for α_2-ANAP.

A similar form of MGSA/gro with non identifiable amino terminus has been isolated from melanomas and has been termed "16 kD moiety of MGSA" (20). The origin of MGSA/gro in psoriatic scales remains speculative at the moment. Although in the epidermis MGSA/gro like immunoreactivity is demonstrable, release of MGSA/gro by keratinocytes has so far not been reported.

Dermal fibroblasts, upon stimulation with IL-1α or TNFα, secrete MGSA/gro (23,29). However, in contrast to α_2-ANAP, amino acids of the amino terminus are alanine and serine. Therefore fibroblasts apparently are not the source of α_2-ANAP.

β1-ANAP ACTIVATES THE PMN-5-LIPOXYGENASE

Extracts obtained from psoriatic scale material led to the production of LTB_4, when PMN were added (data not shown). In order to characterize the 5-lipoxygenase activating principle we performed cation exchange HPLC with proteinaceous contents of the scale extracts. Interestingly investigation of HPLC fractions for LTB_4 synthesis-inducing activity revealed no fraction which elicited LTB_4 production in PMN directly (data not shown). However, when PMN were stimulated with HPLC-fractions in the presence of 10^{-5} M exogenous arachidonic acid a strong peak of LTB_4-producing activity was detected (Figure 7). Purification of this activity (data not shown) led to the conclusion that both β_1- and β_2-ANAP are stimulators of PMN-5-lipoxygenase. However, release of LTB_4 could only be achieved when exogenous arachidonic acid was present. These findings confirm recent studies in our laboratory (30) that NAP-1/IL-8, which is identical with β_2-ANAP, is an activator of PMN-5-lipoxygenase. Moreover, it extends these findings to the 69-residue form of NAP-1/IL-8 (β_1-ANAP). It appears that β_1-ANAP is the by far dominating PMN-5-lipoxygenase activating protein in lesional psoriatic scales.

CONCLUSION

The quantitatively dominant proteinaceous PMN-chemoattractants in psoriatic scales are structurally related to NAP-1/IL-8. The 72-residue form of NAP-1/IL-8 is known to be a powerful PMN-and T-lymphocyte attractant *in vitro* (10,11,1, 4). Although lymphocyte chemotactic activity of the 69 residue form of NAP-1/IL-8 has not been reported, it is likely to be responsible for the major part of cell-specific chemoattraction of both, PMN and lymphocytes into the psoriatic epidermis, because NAP-1/IL-8 does not attract monocytes (10,11) and eosinophils (10) *in vitro*.

Also the presence of LTB_4 in psoriatic lesions may indirectly be related to the presence of PMN- as well as NAP-1/IL-8-related cytokines, because it appears that other cells do not participate similarly in LTB_4 production.

In addition, it is interesting to speculate that part of the epidermal cell hyper-proliferation found in psoriatic patients might be the result of drastically increased concentrations of β_2-ANAP (NAP-1/IL-8) and possibly β_1-ANAP (the 69 residue form of NAP-1/IL-8) because recently NAP-1/IL-8 could be identified as a potent and apparently keratinocyte-specific mitogen for human epidermal cells (31). The finding that, in psoriatic scale material, concentrations of NAP-1/IL-8-related cytokines are up to 150 fold higher than that found in extracts of normal heel callus (19) points towards an important role of NAP-1/IL-8-related cytokines in psoriasis, which might explain the composition of the cellular infiltrate in the psoriatic epidermis as well as in part also epidermal hyperproliferation.

Acknowledgements: This work was supported by a grant from the Deutsche Forschungsgemeinschaft SCHR 305, 1-2. I am indebted to Mrs. Ilse Brandt for editorial help in preparing the manuscript.

REFERENCES

1. Braun-Falco, O., and E. Christophers. 1974. Structural aspects of initial psoriatic lesions. *Arch. Derm. Forsch.* 251: 95-110.
2. Tagami, H., and S. Ofuji. 1977. Characterization of a leukotactic factor derived form psoriatic scale. *Br. J. Dermatol.* 97: 509-518.
3. Brain, S.D., R.D.R. Camp, P.M. Dowd, M.W. Greaves, A.K. Black, A.I. Mallet, and P.M. Woollard. 1983. Leukotriene B_4 and monohydroxy eicosatetraenoic acid-like material are released in biologically active amounts from the lesional skin of patients with psoriasis. *Br. J. Dermatol.* 83: 313-317.

4. Hammarström, S., M. Hamberg, B. Samuelsson, E.A. Duell, M. Stawiski, and J.J. Vorhees. 1975. Increased concentrations of nonsterified arachidonic acid, 12-L-hydroxy-5,8,10,14-eicosatetraenoic acid, prostaglandin E_2 and prostaglandin F_2 in epidermis of psoriasis. *Proc. Natl. Acad. Sci. USA* 72: 5130-5134.
5. Mallet, A.I., F.M. Cunningham, and R. Daniel. 1984. Rapid isocratic high-performance liquid chromatographic purification of platelet activating factor (PAF) and lyso-PAF from human skin. *J. Chromatogr.* 309: 160-164.
6. Fernandez, H.N., and T.E. Hugli. 1976. Partial characterization of human C5a anaphylatoxin. I. Chemical Description of the carbohydrate and polypeptide portion of human C5a. *J. Immunol.* 117: 1688-1693.
7. Tagami, H., Y. Kitano, S. Suehisa, T. Iku, and M. Yamada. 1982. Psoriatic leukotactic factor. Further physicochemical characterization and effect on the epidermal cells. *Arch. Dermatol. Res.* 272: 201-213.
8. Schröder, J.-M., and E. Christophers. 1986. Identification of $C5a_{des\ arg}$ and an Anionic Neutrophil-Activating Peptide (ANAP) in Psoriatic Scales. *J. Invest. Dermatol.* 87: 53-58.
9. Bokisch , V.A., and H.J. Müller-Eberhard. 1970. Anaphylatoxin inactivator of human plasma; its isolation and characterization as a carboxy-peptidase. *J. Clin. Invest.* 49: 2477-2436.
10. Schröder, J.-M., U. Mrowietz, E. Morita, and E. Christophers. 1987. Purification and partial biochemical characterization of a human monocyte-derived neutrophil-activating peptide that lacks Interleukin 1 activity. *J. Immunol.* 139: 3474-3483.
11. Yoshimura, T., K. Matsushima, S. Tanaka, E.E. Robinson, E. Appella, J. J.Oppenheim and E. J. Leonard. 1987. Purification of a human monocyte-derived neutrophil chemotactic factor that shares sequence homology with other host defense cytokines. *Proc. Natl. Acad. Sci. USA* 84: 9233-9237.
12. Walz, A., P. Peveri, H. Aschauer, and M. Baggiolini. 1987. Purification and amino acid sequencing of NAF, a novel neutrophil-activating factor produced by monocytes. *Biochem. Biophys. Res. Commun.* 149: 755-761.
13. Van Damme, J., J. Van Beeumen, G. Opdenakker, and A. Billiau. 1988. NH_2-terminal sequence-characterized human monokine possessing neutrophil chemotactic, skin-reactive and granulocytosis-promoting activity. *J. Exp. Med.* 167: 1364-1376.
14. Gronhoj-Larsen, C., A. O. Anderson, E. Appella, J.J. Oppenheim, and K. Matsushima. 1989. The Neutrophil-Activating Protein (NAP-1) Is Also Chemotactic for T Lymphocytes. *Science* 243: 1464-1466.
15. Westwick, J., S.W. Li, and R.D. Camp. 1989. Novel neutrophil stimulating peptides. *Immunology Today* 10: 146-147.
16. Sauder, D. N., Mounessa, N.L., S.I. Katz, C.A. Dinarello, and J.I. Gallin. 1984. Chemotactic cytokines: the role of leukotactic pyrogen and epidermal cell thymocyte activating factor in neutrophil chemotaxis. *J. Immunol.* 132: 828-832.
17. Luger, T.A., J. A. Charon, M. Colot, M. Micksche, and J.J. Oppenheim. 1983. Chemotactic properties of partially purified human epidermal cell-derived thymocyte-activating factor (ETAF) for polymorphonuclear and mononuclear cells. *J. Immunol.* 131: 816-820.
18. Georgilis, K., C. Schäfer, C.A. Dinarello, and M.S. Klempner. 1987. Human recombinant interleukin 1β has no effect on intracellular calcium or on functional responses of human neutrophils. *J. Immunol.* 139: 3403-3407.
19. Schröder, J.-M., J. Young, H. Gregory, and E. Christophers. 1990. Neutrophil Activating Proteins in Psoriasis. Submitted.
20. Richmond, A., E. Balentien, H.G. Thomas, G. Flaggs, D.E. Barton, J. Spiess, R. Bardoni, U. Francke, and R. Derynck. 1988. Molecular characterization and chromosomal mapping of melanoma growth stimulatory activity, a growth factor structurally related to β-thromboglobulin. *EMBO J.* 7: 2025-2033.
21. Kähler, S., E. Christophers, and J.-M. Schröder. 1988. Plasma lactoferrin reflects neutrophil activation in psoriasis. *Br. J. Dermatol.* 119: 289-293.
22. Schröder, J.-M., and E. Christophers. 1989. Secretion of novel and homologues neutrophil activating peptides by LPS-stimulated human endothelial cells. *J. Immunol.* 142: 244-251.
23. Schröder, J.-M., M. Sticherling, H.-H. Henneicke, W.C. Preissner, and E. Christophers. 1990. IL-1α or TNFα stimulate release of three IL-8-related neutrophil chemotactic proteins in human dermal fibroblasts. *J. Immunol.* 144: 2223-2232.
24. Barker, J.N. W.N., V. Sana, R.S. Mitra, V.M. Dixit, and B.J. Nickoloff. 1990. Marked Synergism between Tumor Necrosis Factor α and Interferon in Regulation of Keratinocyte-derived Adhesion Molecules and Chemotactic Factors. *J. Clin. Invest.* 85: 605-608.

25. Barker, J.N. W.N., C.E.M. Griffiths, R.S. Mitra, J. T. Elder, V. Dixit, S.Kunkel, and B.J. Nickoloff. 1990. Keratinocyte-Derived Interleukin-8 (IL-8): Regulation by TPA and Urushiol and Detection in Inflamed Skin. *J. Invest. Dermatol.* In press.
26. Matsushima, K. and J.J. Oppenheim. 1989. Interleukin 8 and MCAF: Novel Inflammatory Cytokines Inducible by IL 1 and TNF. *Cytokine* 1: 2-13.
27. Baggiolini, M., A. Walz, and S.L. Kunkel. 1989. Neutrophil-activating Peptide-1/Interleukin 8, a Novel Cytokine That Activates Neutrophils. *J. Clin. Invest.* **84**: 1045-1049.
28. Schröder, J.-M., N. Persoon, and E. Christophers. 1990. Lipopolysaccharide-stimulated human monocytes secrete apart from NAP-1/IL-8 a second neutrophil-activating protein: NH_2-terminal amino acid sequence-identity with melanoma growth stimulatory activity (MGSA/gro). *J. Exp. Med.* 171: 1091-1100.
29. Golds, E.E., P.P. Mason, and P. Nyirkos. 1990. Inflammatory cytokines induce synthesis and secretion of gro protein and a neutrophil chemotactic factor but not β_2-microglobulin in human synovial cells and fibroblasts. Biochem. J. 259: 585-588.
30. Schröder, J.-M. 1989. The monocyte-derived neutrophil activating peptide (NAP/Interleukin 8) stimulates human neutrophil arachidonate 5-lipoxygenase but not release of cellular arachidonate. J. Exp. Med. 170: 847-863.
31. Krueger, G, C. Jörgensen, C. Miller, J.-M. Schröder, M. Sticherling, and E. Christophers. 1990. Effects of IL-8 on epidermal proliferation. J. Invest. Dermatol. 94: 545 (abstract).
32. Schägger, H., and G. von Jagow. 1987. Tricine-Sodium, Dodecyl Sulfate-Polyacrylamide Gel Electrophoresis for the Separation of Proteins in the Range from 1 to 100 kDa. Anal. Biochem. 166: 368-377.

CHEMOTACTIC CYTOKINES IN INFLAMMATORY SKIN DISEASE

Richard Camp, Kevin Bacon, Nicholas Fincham, Kay Mistry, Janet Ross, Frances Lawlor, Daniel Quinn, and Andrew Gearing[1]

Institute of Dermatology, St. Thomas's Hospital, London SE1 7EH
[1]British Biotechnology, Watlington Road, Oxford OX4 5LY, UK

INTRODUCTION

The production of cytokines in human skin *in vivo* has been investigated and found to be more selective than studies *in vitro* have suggested. Thus, stratum corneum samples from the skin lesions of psoriasis contain interleukin 8 (IL-8)-like material, but assay for a range of other compounds suggested that no other defined, biologically active cytokine was present in increased levels when compared with those in control heel stratum corneum samples. Selective cytokine release was also found on analysis of chamber fluid samples from the skin lesions of the cutaneous T-cell lymphoma, mycosis fungoides. Assay of normal epidermal samples has shown the presence of biologically active amounts of IL-1 like material, chromatographic purification and the use of neutralizing antibodies indicating the presence of IL-1α but negligible IL-1β activity in normal heel stratum corneum extracts. No other biologically active cytokine has been detected in normal skin. The IL-1α-like material recoverable from normal human epidermis possesses potent inflammatory properties when injected intradermally, but appears not to be biologically available under normal *in vivo* conditions, possibly through intracellular retention in keratinocytes, membrane association or control by an inhibitor. The release of preformed IL-1 following membrane perturbation or other events may constitute a primary mechanism for the induction of inflammation in human skin. For reasons to be outlined below, IL-1 may be less important in the maintenance of chronic inflammatory changes, at least in psoriasis, IL-8 possibly playing a more significant role.

PSORIASIS

This common, idiopathic chronic skin disease is characterised by inflamed, scaling, skin lesions containing infiltrates of neutrophils, lymphocytes and monocytes. A range of leukocyte attractants has been identified in samples from psoriatic skin lesions, including leukotriene B_4 (1,2), 12[R]-hydroxyeicosatetraenoic acid (3), platelet activating factor (4) and $C5a_{des\ arg}$ (5). In 1986 a novel neutrophil activating and attractant peptide distinct from $C5a_{des\ arg}$ but chromatographically similar to IL-1 was identified in aqueous extracts of the scales obtained from the

Chemotactic Cytokines, Edited by J. Westwick *et al.*
Plenum Press, New York, 1991

surface of psoriatic lesions (5,6). Subsequently, successive reversed phase and anion exchange high performance liquid chromatography (HPLC) purification of this novel attractant material revealed the presence of several biologically active components (7). Recently, three components were separated by anion exchange HPLC, and shown to induce dilution-related activity in an agarose microdroplet neutrophil migration assay. Each component was neutralised by a sheep IL-8 antiserum (8). Preliminary sequence analysis has confirmed the identity of at least one component as an IL-8 species, a second component showing homology with the amino-terminal sequence of melanoma growth-stimulating activity (9).

Aqueous extracts of psoriatic lesional scale have also been analyzed for a range of other cytokines, by using biological and immunological assays and the levels compared with those in aqueous extracts of normal heel stratum corneum, as control (8). There was no detectable IL-2 (CTLL proliferation assay), IL-4 (two-site ELISA), IL-6 (B9 proliferation assay) or granulocyte colony stimulating factor (GCSF; two-site immunoradiometric assay) in any sample. Granulocyte-macrophage colony-stimulating factor (GMCSF; two-site ELISA) immunoreactivity was detectable in trace amounts but was not elevated in lesional samples. Immunoassays revealed increased levels of interferon (IFN) α and gamma, and tumour necrosis factor (TNF) α and β, in lesional samples, but bioassays for IFN (viral cytopathic effect reduction assay) and TNF (L929 cytotoxicity assay) activities were negative for all samples. Substantial quantities of IL-1 activity (EL-4 NOB-1 bioassay) were present in normal heel stratum corneum extracts (see section on IL-1 in normal skin, below), but levels of activity were grossly reduced in the psoriatic lesional stratum corneum extracts (2143 ± 354 U ml^{-1}, mean ± s.e. mean for normal heel stratum corneum, n=5; 52 ± 22 U ml^{-1} for psoriatic lesional stratum corneum, n=7; supernatants prepared by homogenising 70 mg stratum corneum in 4 ml PBS in each case) (8). Thus, of the nine compounds assayed, IL-8 was the only biologically active cytokine which was present in greater amounts in lesional than control samples. Aqueous psoriatic scale extracts induced dilution-related responses in the agarose microdroplet neutrophil migration assay. These were largely (but not completely) neutralised by IL-8 antiserum, a finding which suggests that a major portion of the neutrophil attractant activity in these samples is due to IL-8-like material. Furthermore, little or no neutrophil attractant activity was found in aqueous extracts of normal heel stratum corneum (8).

The failure to demonstrate IL-6 activity in psoriatic stratum corneum extracts by the B9 proliferation assay contrasts with the report (10) that IL-6 is highly expressed in psoriatic epidermis, as determined by immunohistochemistry and in situ hybridisation. A second, preliminary report described only low, but detectable, IL-6 activity in lesional suction blister fluid samples as compared with no detectable activity in control samples, by use of the B9 proliferation assay (11). In view of the above evidence for the presence of immunoreactive IFN and TGF species but a lack of corresponding biological activity in psoriatic samples (8), we believe that more work is needed to show that the substantial IL-6 immunoreactivity demonstrated in psoriatic lesions (10) is due to a biologically active compound. The presence of IL-6 biological activity in psoriatic sera may be due to the release of active material from the skin (10), but does not prove that the highly-expressed IL-6 immunoreactive species in psoriatic lesions (10) is biologically active.

The above findings point to IL-8 as a potentially important mediator of the pathology of psoriasis. Further evidence as to its potential importance derives from the findings that IL-8 is also a potent *in vitro* lymphocyte attractant (12,13) and keratinocyte growth factor (14), and that IL-8 messenger RNA and/or peptide may be produced by keratinocytes (15,16) and dermal fibroblasts (16) *in vitro*. However, preliminary evidence suggests that its production is not specific to psoriasis but may occur in other inflammatory skin diseases (18). As indicated in the opening paragraph of this section, IL-8 is also not the only neutrophil attractant present in psoriatic lesions. Further work is required to clarify the relative amount of biological activity generated by the peptide and lipid leukocyte attractant components in psoriatic lesions.

IL-8 has been found to induce human lymphocyte migration *in vitro* by at least three different groups (12,13,19). We have found IL-8-induced lymphocyte migration to be potently inhibited by voltage-dependent calcium channel antagonists (13,20). IL-8 induced responses were also dependent on protein kinase C (PKC) activation as shown by the inhibitory effects of the specific PKC inhibitors, Ro 31-8220 and Ro 31-7549 (21). The involvement of GTP-binding proteins in these IL-8-induced responses was also suggested by the inhibitory effects of cholera and pertussis toxins (21). There is however, controversy concerning the ability of IL-8 to induce lymphocyte infiltrates *in vivo*. One group has reported the presence of lymphocytes in guinea pig skin following intradermal injection of low doses of IL-8 (12), but another failed to demonstrate lymphocyte infiltration after injection of IL-8 into human skin (22). Clearly, the importance of IL-8 in psoriasis and other inflammatory diseases will depend on its ability to induce relevant changes in human skin *in vivo*. More work is needed to clarify this issue.

We have recently described the presence of >10kD human monocyte attractant activity in psoriatic lesional stratum corneum samples, as determined by ultrafiltration of supernatants on YM10 membranes and analysis by an *in vitro* migration assay involving 48-well microchemotaxis chambers. HPLC purification indicated the presence of at least two biologically active compounds (23). Further work is required to determine the relationship of this material to recently described monocyte attractant cytokines.

CUTANEOUS T-CELL LYMPHOMA

In the cutaneous T-cell lymphoma, mycosis fungoides, there is, in addition to the infiltration of malignant T-cells, a mixed leukocyte infiltrate, especially in the earlier stages. This could be induced by locally released cytokines, which may play an important role in the evolution of the disease. We sampled lesional and clinically uninvolved skin by using a skin chamber technique in which acrylic cylinders are fixed to abraded areas with cyanoacrylate adhesive. PBS (1 ml) is placed in each chamber, then discarded after 2 min, serving as a wash. A further 1 ml PBS is added to each chamber, removed after 60 min and assayed for the presence of cytokines as indicated in Table 1 (24).

Table 1. Cytokine levels in chamber fluid samples from lesions and clinically uninvolved skin in mycosis fungoides

Cytokine	Assay	Detection limit ($pg\ ml^{-1}$)	Chamber fluid levels ($pg\ ml^{-1}$)[1] uninvolved	lesional
IL-1	EL-4 NOB-1	0.2	627 ± 273	208 ± 78[2]
IL-6	B9	0.5	22 ± 9	353 ± 137[3]
TNF	L929	20	0	0
IL-2	CTLL	500	0	0
IFN gamma	IRMA	20	0	0
GMCSF	ELISA	10	0	0

[1] mean ± s.e. mean values given, n=7
[2] $p = 0.05$ and [3] $p < 0.05$ (Wilcoxon signed rank test)

As for normal heel stratum corneum, chamber fluid from clinically uninvolved skin contained substantial, although variable, quantities of IL-1 activity, and levels were reduced in lesional samples (see section on IL-1 in normal skin, below). IL-6-like material was the only cytokine, of those tested, which was found to be present in increased amounts in lesional samples. Although the presence of other cytokines, such as IL-8, cannot be excluded, this and the psoriasis data support the possibility that cytokine production in human skin disease is regulated and not characterised by the uncontrolled production of a "cascade" of compounds.

There have been two previous reports of IL-1 production in mycosis fungoides. In one report, similar levels of IL-1 activity were found in chamber fluid samples from lesional and clinically uninvolved skin of three patients (25). In this case the relatively non-specific murine thymocyte co-stimulator assay, which may not distinguish IL-1 from IL-6, was used. The reduced IL-1 but raised IL-6 activity in lesions, and the higher IL-1 and lower IL-6 activity in samples from uninvolved skin, might generate thymocyte co-stimulatory activity which does not differ significantly in the samples from the two sources. In the second report raised levels of IL-1β were found in lesional samples, as determined by immunohistological analysis of skin biopsies (26). As outlined in the following section dealing with IL-1, we found that IL-1α-like material is present in samples form normal skin and that little or no biologically active IL-1β may be recovered. Therefore, the use of an antibody directed only against IL-1β (26) may not reflect the overall production of IL-1.

The biological effects of locally released IL-6 in mycosis fungoides lesions are uncertain. However, we have now shown that IL-6 induces human lymphocyte migration *in vitro*, affecting both $CD4^+$ and $CD8^+$ T-cell subpopulations (27). IL-6 may therefore play a role in inducing cutaneous T-cell infiltrates in mycosis fungoides.

INTERLEUKIN 1 IN NORMAL SKIN

Substantial quantities of IL-1 activity may be extracted from normal human epidermis. In the first report of this phenomenon, 1 g normal heel stratum corneum was found to contain as much IL-1 activity as several litres of monocyte supernatant (28). Subsequently, human epidermal samples obtained by suction blister, keratome slicing and heat separation techniques were found to contain biologically active IL-1-like material (29). We have confirmed these findings by analysis of normal heel stratum corneum and chamber fluid from abraded normal skin (7,30). IL-1 has, however, been shown to induce potent inflammatory reactions in mouse footpad (31), rat ear (32) and rabbit skin (33), and intradermal injection of femtomole amounts (10-100 U) of recombinant IL-1α in human volunteers induced florid, prolonged erythematous reactions containing mixed leukocyte infiltrates (34). We therefore questioned whether the IL-1-like material in human epidermis possesses the same inflammatory properties as recombinant IL-1, particularly as the skin-derived material has never been subjected to amino acid sequence analysis.

Ultrafiltration fractions (>10kD) of normal heel stratum corneum were therefore subjected to reversed phase HPLC. A portion of each 1 min fraction was tested for IL-1 activity and a further portion evaporated, reconstituted in sterile PBS and injected intradermally into the arms of the corresponding heel stratum corneum donor. There was close co-elution of inflammatory activity, recorded as area of erythema, and IL-1 activity, measured in an EL-4 NOB-1 bioassay, and no fractions other than those containing IL-1 activity induced inflammatory reactions. The active fractions, eluting at about 50% acetonitrile, were pooled and repurified by anion exchange HPLC with pH gradient elution. Each 1 min fraction was tested for IL-1 activity and for inflammatory activity as before. Again, inflammatory and IL-1 activity co-eluted, and had the same retention time as a pI 5 marker, little or no activity of either type co-eluting with a pI 7 marker. These results, which were consistent in the three subjects tested, indicate that inflammatory quantities of pI 5 IL-1α-like material are present in normal heel stratum corneum samples. Activity induced by heel stratum corneum extracts in the EL-4 NOB-1 bioassay was neutralised by a sheep IL-1α antiserum but not by an IL-1β antiserum, confirming that no biologically active IL-1β is recoverable from this source (30). To investigate whether this phenomenon is confined to heel stratum corneum, skin chamber fluid samples from the thighs of two normal volunteers were subjected to similar analysis. Two circular areas (2 cm diameter) on the thighs were abraded by repeated application and removal of adhesive tape and two acrylic cylinders applied. PBS (1 ml per chamber) was added, removed after 30 min, pooled, concentrated on YM10 membranes and purified by anion exchange HPLC. Analysis of 1 min fractions for IL-1 and inflammatory activity, as above, confirmed that inflammatory quantities of IL-1α-like material were released from adhesive tape-stripped normal skin into the chamber fluid. The short contact period (30 min) suggested the release of pre-formed material into chamber fluid rather than recovery of newly-synthesised IL-1 (30). This confirmed the previous demonstration (29) that biologically active IL-1 may be recovered from skin sites other than the heel, and that inflammatory amounts are readily released from localised areas of tape-stripped skin.

The source of this normal epidermal IL-1 may be keratinocytes which, in culture, have been shown to produce messenger RNA for IL-1, particularly IL-1α (35). IL-1 has also been recovered from sweat (36) which may contribute to the activity in normal heel stratum corneum but seems a less likely source of the IL-1 activity released into skin chamber fluid.

Intradermal injection of IL-1α-like material (60 U) purified from heel stratum corneum by reversed phase and anion exchange HPLC caused inflammatory reactions of a similar magnitude and duration to those seen with recombinant IL-1α (30), erythema appearing after 1-2 hours, reaching a maximum at approximately 9 hours, and lasting at least 24 hours. Biopsies 4 hours after intradermal injection of 60 U autologous IL-1α-like material showed mixed infiltrates of neutrophils, monocytes and lymphocytes (Table 2) (30).

Table 2. Leukocyte subpopulations in skin biopsies 4 hours after intradermal injection of 60 U autologous IL-1α-like material from heel stratum corneum

Leukocyte type and staining method	Cell numbers[1]	
	PBS[2]	IL-1
neutrophils (chloracetate esterase)	1 ± 1	49 ± 15[3]
monocyte/macrophages (OKM5)	95 ± 15	284 ± 42[3]
CD4 positive T-cells (Leu 3a)	97 ± 8	167 ± 25[3]
CD8 positive T-cells (Leu 2a)	20 ± 3	25 ± 5

[1]mean ± s.e. mean number of cells per 5 high power fields immediately below the epidermis; n=3
[2]PBS injected intradermally as control
[3]$P < 0.05$, Mann-Whitney U-test, on comparison with PBS injection sites

Two authoritative articles conclude that IL-1 does not induce neutrophil migration *in vitro* (37,38), although a further two reports claim the opposite (39,40), one (40) demonstrating that IL-1-induced neutrophil migration *in vitro* is albumin dependent. IL-1 receptors have also been demonstrated on neutrophils (41). However, we have found that normal heel stratum corneum extracts induce little or no *in vitro* neutrophil migration, in spite of the presence of substantial levels of IL-1 activity (7,8). It is therefore possible that the IL-1-induced neutrophil infiltrates demonstrable after intradermal injection, as described above, are due to the rapid induction of the synthesis of attractants such as IL-8 by keratinocytes (15,16) or fibroblasts (17). The substantial monocyte infiltrates induced by intradermal injection of the IL-1α-like material are also likely to be due to induction of the synthesis of other substance(s), as IL-1 has not been found to be a monocyte attractant *in vitro* (D.G. Quinn, unpublished observations). The $CD4^+$ T-cell infiltrates may, however, be due to a direct effect, as we (13,42) and others (43,44) have found IL-1 to induce lymphocyte migration *in vitro*. By immunocytochemical analysis of lymphocytes adherent to the undersurface of polycarbonate filters following a migration assay in which mixed human peripheral blood lymphocytes were used, we have shown that recombinant IL-1α induces the migration of $CD4^+$ but not $CD8^+$ T-cells (J.S. Ross., K.B. Bacon and R.D.R. Camp, submitted for publication). This correlates well with the finding of increased $CD4^+$ T-cell infiltrates but not $CD8^+$ cells following intradermal injection of the IL-1α-like material as described above.

The erythema following IL-1 injection is presumably an affect of vasodilation, which may be due to the local release of vasoactive eicosanoids (45), platelet activating factor (46), or possibly to a direct but protein synthesis-dependent effect on vascular smooth muscle (47).

The finding of biologically active IL-1-like material in epidermal samples from normal, non-inflamed human skin suggests that IL-1 is not biologically available under physiological conditions. This may be due to intracellular retention, membrane association or control by an inhibitor, although evidence for an IL-1 inhibitor in normal skin has not been reported. We postulate that the release of preformed IL-1 following membrane perturbation, such as may occur after wounding or in disease, constitutes a primary mechanism for the induction of inflammation in human skin. The reduced levels of IL-1 activity in samples from psoriasis (7,8) and mycosis fungoides (24) lesions possibly indicate the release of preformed material and resulting depletion of lesional levels. This suggests that IL-1 may be more important in the initiation of inflammatory skin lesions rather than in the maintenance of chronicity, cytokines such as IL-6 and IL-8 possibly playing a more important role in the latter process. Finally, low level release of IL-1 from normal epidermis may, through its lymphocyte attractant properties, be relevant to the induction of physiological lymphocyte trafficking in skin.

CYTOKINE-INDUCED MONONUCLEAR LEUKOCYTE MIGRATION

We have developed an *in vitro* lymphocyte migration assay which incorporates mixed peripheral blood lymphocytes (PBL) purified by density gradient centrifugation and two plastic adherence steps, the second consisting of an 18 hour incubation in medium containing foetal calf serum (48). Migration of these cells ($<1\%$ monocytes) is measured in a 48-well "micro-chemotaxis" chamber with 8 μm pore-size polyvinylpyrrolidone-free polycarbonate filters. Cells (50 μl of 2 x 10^6 ml^{-1} suspension) are placed in the upper chambers and test material (25μl) in the lower wells. After a 60 min incubation in the absence of serum, the number of cells that have migrated to the filter undersurface is quantified by image analysis (48).

Although PBL are less adherent to polycarbonate filters than other leukocytes, concentration-related responses to a range of agonists may be obtained in this assay if the above in vitro conditions are strictly employed. That the assay measures active PBL migration rather than adherence of passively falling cells is evidenced by the fact that concentration-related adherence to the lower filter surface is seen when agonist is placed in the lower wells only, but not when agonist is placed in both upper and lower chambers according to a modification of the checkerboard system of Zigmond and Hirsch (48,49,50). Furthermore, although small numbers of cells do drop off the filter into lower wells, the number does not reciprocally decrease with the increase in the number of cells adherent to the lower filter surface as agonist concentrations are increased (48). In addition, responses are inhibited when the assay is carried out at 4^0 C or in the presence of cytochalasin B, which interferes with cellular microfilament integrity (27).

A range of cytokines has been found to induce dilution related responses in this PBL migration assay, including IL-1α, IL-1β, IL-2 (13,42), IL-3, IL-4, IL-6 (27; Figure 1), IL-7 (K.B. Bacon, unpublished observations) and IL-8 (13). Interferon gamma induced only minimal responses (Figure 1), while transforming growth factor β, granulocyte colony-stimulating factor and macrophage colony-stimulating factor were inactive (27).

Thus, a wide range of cytokines is capable of inducing *in vitro* PBL migration. The *in vitro* lymphocyte attractant effects of IL-1 (43,44), IL-2 (51,52) and IL-8 (12,19) have been reported by others, but the responses to the other cytokines described above are, to our knowledge, novel findings. Immunophenotyping of lymphocytes adherent to the undersurface of polycarbonate filters in response to selected cytokines has indicated that IL-1, IL-3 and IL-4 specifically induce the migration of $CD4^+$ cells while IL-6 and IL-8 induce migration of both $CD4^+$ and $CD8^+$ cells (24,27; J.S. Ross., K.B. Bacon & R.D.R. Camp, submitted for publication).

In contrast to the lymphocyte responses, transforming growth factor β induced concentration-related human monocyte migration, whereas IL-1, IL-3, IL-4, IL-6 and IL-8 were inactive, and the effects of interferon gamma were inconsistent (D.G. Quinn, unpublished observations). These differences in the selectivity of the

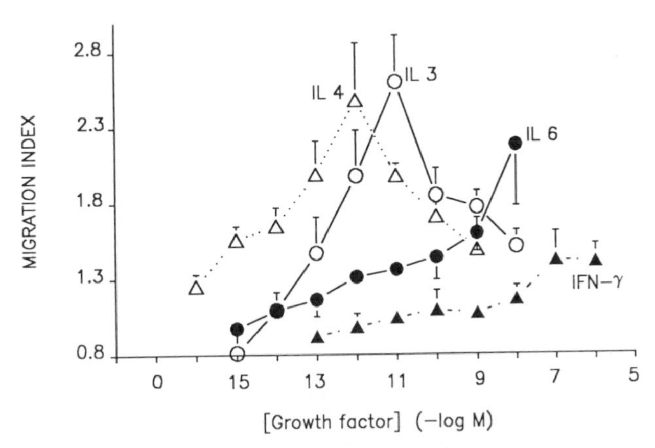

Figure 1. Dilution-related lymphocyte migration in response to a range of cytokines. Mean ± s.e. migration index values are shown. The migration index represents the ratio of response in the presence of agonist to response in the presence of medium alone.

locomotor responses of lymphocytes and monocytes to cytokines may be of pathophysiological significance.

Acknowledgements: The financial support of the Medical Research Council is gratefully acknowledged.

REFERENCES

1. Brain, S.D., R.D.R. Camp, F.M. Cunningham, P.M. Dowd, M.W. Greaves, and A. Kobza Black. 1984. Leukotriene B_4-like material in scale of psoriatic skin lesions. *Br. J. Pharmacol.* **83**: 313-317.
2. Brain, S.D., R.D.R. Camp, P.M. Dowd, A. Kobza Black, and M.W. Greaves. 1984. The release of leukotriene B_4-like material in biologically active amounts from the lesional skin of patients with psoriasis. *J. Invest. Dermatol.* **83**: 70-73.
3. Woollard, P.M. 1986. Stereochemical difference between 12-hydroxy-5,8,10,14-eicosatetraenoic acid in platelets and psoriatic lesions. Biochem. Biophys. Res. Commun. **136**: 169-176.
4. Mallet, A.I., and F.M. Cunningham. 1985. Structural identification of platelet activating factor in psoriatic scale. *Biochem. Biophys. Res. Commun.* **126**: 192-198.
5. Schröder, J.M., and E. Christophers. 1986. Identification of $C5a_{des\ arg}$ and an anionic neutrophil activating peptide (ANAP) in psoriatic scales. *J. Invest. Dermatol.* **87**: 53-58.
6. Camp, R.D.R., N.J. Fincham, F.M. Cunningham, M.W.Greaves, J. Morris, and A.C. Chu. 1986. Psoriasis skin lesions contain biologically active amounts of an interleukin 1-like compound. *J. Immunol.* **137**: 3469-3474.
7. Fincham, N.J., R.D.R. Camp, A.J.H. Gearing, C.R. Bird, and F.M. Cunningham. 1988. Neutrophil chemoattractant and IL-1-like activity in samples from psoriatic skin lesions. Further characterisation. *J. Immunol.* **140**: 4294-4299.
8. Gearing, A.J.H., N.J. Fincham, C.R. Bird, A.M. Wadhwa, A. Meager, J.E. Cartwright, and Camp R.D.R. 1990. Cytokines in skin lesions of psoriasis. *Cytokine*. **2**: 68-75.
9. Schröder, J.M., and E. Christophers. 1989. Amino acid sequence characterization of two ultrastructurally related neutrophil activating peptides obtained from lesional psoriatic scales. *J. Invest. Dermatol.* **92**: 515(abstract).
10. Grossman, R,M., J. Krueger, D. Yourish, A. Granelli-Piperno, D.P. Murphy, L.T. May, T.S. Kupper, P.B. Seghal, A.B. Gottlieb. 1989. Interleukin 6 is expressed in high levels in psoriatic skin and stimulates proliferation of cultured human keratinocytes. *Proc. Natl. Acad. Sci. USA.* **86**: 6367-6371.
11. Prens E.P., J. van Damme, M. Bakkus, K. Brakel, R. Benner, and T. van Joost. 1989. IL-1 and IL-6 in psoriasis. *J. invest. Dermatol.* **93**: 570 (abstract).
12. Larsen, C.G., A.O. Anderson, E. Appella, J.J. Oppenheim, and K. Matsushima, 1989. The neutrophil-activating protein (NAP-1) is also chemotactic for T lymphocytes. *Science.* **243**: 1464-1466.
13. Bacon, K.B., J. Westwick, and R.D.R. Camp. 1989. Potent and specific inhibition of IL-8-, IL-1α- and IL-1β-induced lymphocyte migration by calcium channel antagonists. *Biochem. Biophys. Res. Commun.* **165**: 349-354.
14. Krueger, G., C. Jorgensen, C. Miller, J. Schröder, M. Sticherling, and E. Christophers. 1990. Effects of IL-8 on epidermal proliferation. *J. Invest. Dermatol.* **94**: 545 (abstract).
15. Barker, J.N.W.N., V. Sharma, R.S. Mitra, V.M. Dixit, and B.J. Nickoloff. 1990. Marked synergism between tumour necrosis factor-α and interferon-in regulating keratinocyte-derived adhesion molecules and chemotactic factors. *J. Clin. Invest.* **85**: 606-608.
16. Barker, J.N.W.N., M.L. Jones, C. Swenson, R.S. Mitra, J.T.Elder, J.C. Fantone, P.A. Ward, V.M. Dixit, and B.J. Nickoloff. 1990. Keratinocyte (KC) production of a biologically active monocyte chemoattractant. *J. Invest. Dermatol.* **94**: 505 (abstract).
17. Schröder, J.M., M. Sticherling, H.H. Heneicke, W.C. Preissner, and E. Christophers, 1990. IL-1 or tumour necrosis factor-α stimulate release of three NAP-1/IL-8-related neutrophil chemotactic proteins in human dermal fibroblasts. *J.Immunol.* **144**: 2223-2232.
18. Schröder, J.M., and E. Christophers. 1990. Determination of IL-8-related cytokines in the stratum corneum of chronic as well as acute inflammatory skin diseases. *J. Invest. Dermatol.* (abstract in press).
19. Leonard E.J., A. Skeel, T. Yoshimura, K.Noer, S. Kutvirt, and D. van Epps. 1990. Leukocyte specificity and binding of human neutrophil attractant/activation protein-1. *J.Immunol.* **144**: 1323-1330.

20. Bacon, K.B., and R.D.R. Camp. 1989. Calcium channel antagonists inhibit leukotriene (LT)B_4 and interleukin(IL) 1α-induced lymphocyte migration. *Skin Pharmacol.* 2: 47 (abstract).
21. Bacon, K.B., and R.D.R. Camp. 1990. Interleukin (IL)-8-induced *in vitro* human lymphocyte migration is inhibited by cholera and pertussis toxins and inhibitors of protein kinase C. *Biochem. Biophys. Res. Commun.*, in press.
22. Leonard, E., T. Yoshimura, S. Tanaka, and M. Raffeld. 1989. Neutrophil infiltration caused by intradermal injection of neutrophil attractant protein-1 (NAP-1/IL-8)into human skin. *Cytokine* 1: 151 (abstract).
23. Quinn, D.G., and R.D.R. Camp. 1990. Novel monocyte attractants in stratum corneum from psoriatic lesions. *J. Invest. Dermatol.* 94: 569 (abstract).
24. Lawlor, F., N.P. Smith, R.D.R. Camp, K.B. Bacon, A. Kobza-Black, M.W. Greaves, and A.J.H. Gearing. 1990. Skin exudate levels of interleukin 6, interleukin-1, and other cytokines in mycosis fungoides. *Br. J. Dermatol.* in press.
25. Dowd, P., B.A. Hudspith, R.M. Barr, and J.A. Miller. 1987. Release *in vivo* of interleukin-1 (IL-1)-like activity by human skin. *Clin. Exp. Immunol.* 67: 608-610.
26. Tron, V.A., D. Rosenthal, and D.N. Sauder. 1988. Epidermal interleukin-1 is increased in cutaneous T-cell lymphoma. *J. Invest. Dermatol.* 90: 373-381.
27. Bacon, K., A. Gearing, and R. Camp. 1990. Induction of *in vitro* human lymphocyte migration by interleukin 3, interleukin 4, and interleukin 6. *Cytokine*. 2: 100-105.
28. Gahring, L.C., A. Buckley, and R.A. Daynes. 1985.Presence of epidermal-derived thymocyte activating factor/interleukin 1 in normal human stratum corneum. *J. Clin. Invest.* 76: 1585-1591.
29. Hauser, G., J.H. Saurat, A. Schmitt, F. Jaunin and J.M. Dayer. 1986. Interleukin 1 is present in normal human epidermis. *J. Immunol.* 136: 3317-3323.
30. Camp, R., N. Fincham, J. Ross, C. Bird, and A. Gearing.1990. Potent inflammatory properties in human skin of interleukin-1α-like material isolated from normal skin. *J. Invest. Dermatol.* 94: 735-741.
31. Granstein, R.D., R. Margolis and S.B. Mizel. 1986. *In vivo* inflammatory activity of epidermal cell-derived thymocyte activating factor and recombinant interleukin 1 in the mouse. *J. Clin. Invest.* 77: 1020-1027.
32. De Young, L.M., D.A. Spires, J. Kheifets, and T.G. Terrell. 1987. Biology and pharmacology of recombinant human interleukin-1β-induced rat ear inflammation. *Agents and Actions*. 21: 325-327.
33. Rampart, M., and T.J. Williams. 1988. Evidence that neutrophil accumulation induced by interleukin-1 requires both local protein synthesis and neutrophil CD18 antigen expression *in vitro*. *Br. J. Pharmacol.* 94: 1143-1148.
34. Dowd, P.M., R.D.R. Camp, and M.W. Greaves. 1988. Human recombinant interleukin-1α is pro-inflammatory in normal skin. *Skin Pharmacol.* 1: 30-37.
35. Kupper, T.S., D.W. Ballard, A.O. Chua, J.S. McGuire, P.M.Flood, M.W. Horowitz, R. Langdon, L. Lightfoot, and U. Gubler. 1986. Human keratinocytes contain mRNA indistinguishable from monocyte interleukin 1α and β mRNA. *J. Exp. Med.* 164: 2095-2100.
36. Reitamo, S,. H.S.I. Antilla, L. Didierjean, J-H. Saurat. 1990. Immunohistochemical identification of interleukin 1α and β in human eccrine sweat-gland apparatus. *Br. J. Dermatol.* 122: 315-323.
37. Georgilis, K., C. Schaefer, C.A. Dinarello. and M.S. Klempner. 1987. Human recombinant interleukin 1β has no effect on intracellular calcium or on functional responses of human neutrophils. *J. Immunol.* 138: 3403-3407.
38. Yoshimura, T., K. Matsushima, J.J. Oppenheim, and E.J. Leonard. 1987. Neutrophil chemotactic factor produced by lipopolysaccharide (LPS)-stimulated human blood mononuclear leukocytes: partial characterisation and separation from interleukin 1 (IL-1). *J. Immunol.* 139: 12788-793.
39. Westmacott, D., J. Wadsworth, and D.P. Bloxham. 1987. Chemotactic activity of recombinant human interleukin 1. *Agents and Actions*. 21: 323-324.
40. Maloff, B.L., J.E. Shaw, and D. Fox. 1988. A chemotaxis assay using human polymorphonuclear leukocytes stimulated by IL-1. *J. Immunol.* Methods. 112: 145-146.
41. Parker, K.P., W.R. Benjamin, K.L. Kafka, and P.L. Kilian. 1989. Presence of IL-1 receptors on human and murine neutrophils. Relevance to IL-1-mediated affects in inflammation. *J. Immunol.* 142: 537-542.
42. Ross, J.S., K.B. Bacon, and R.D.R. Camp. 1990. Potent and selective inhibition of *in vitro* lymphocyte migration by cyclosporin and dexamethasone. *Immunopharmacol. Immunotoxicol.* 12: in press.
43. Hunninghake, G.W., A.J. Glazier, M.M. Monick, and C.A.Dinarello. 1987. Interleukin 1 is a chemotactic factor for human T-lymphocytes. *Am. Rev. Resp. Dis.* 135: 66-71.

44. Miossec, P., C-L. Yu, and M. Ziff. 1984. Lymphocyte chemotactic activity of human interleukin 1. *J. Immunol.* 133: 2007-2011.
45. Rossi, V., F. Brevario, P. Ghezzi, E. Dejana, and A. Mantovani. 1985. Prostacyclin synthesis induced in vascular cells by interleukin 1. *Science.* 229: 174-176.
46. Bussolini, F., F. Brevario, C. Tetta, M. Aglietta, A. Mantovani, and E. Dejana. 1986. Interleukin 1 stimulates platelet-activating factor production in cultured human endothelial cells. *J. Clin. Invest.* 77: 2027-2033.
47. Beasley, D., R.A. Cohen, and N.G. Levinsky. 1989. Interleukin 1 inhibits contraction of vascular smooth muscle. *J. Clin. Invest.* 83: 331-335.
48. Bacon, K.B., R.D.R. Camp, F.M. Cunningham, and P.M. Woollard. 1988. Contrasting *in vitro* chemotactic activity of the hydroxyl enantiomers of 12-hydroxy-5,8,10,14-eicosatetraenoic acid. *Br. J. Pharmacol.* 95: 966-975.
49. Bacon, K.B., N.J. Fincham, and R.D.R. Camp. 1990. Stimulation of lymphocyte migration by a novel low molecular weight compound in normal skin and plasma. *Eur. J. Immunol.* 20: 565-571.
50. Zigmond, S.H., and J.G. Hirsch. 1973. Leukocyte locomotion and chemotaxis. *J. Exp. Med.* 137: 387-410.
51. Kornfeld, H., J.S. Berman, D.J. Beer. and D.M. Center.1985. Induction of human T lymphocyte motility by interleukin 2. *J. Immunol.* 137: 3887-3890.
52. Robbins, R.A., L. Klassen, J. Rasmussen, M.E.M. Clayton, and W.D. Russ. 1986. Interleukin-2-induced chemotaxis of human T-lymphocytes. *J. Lab. Clin. Med.* 108: 340-345.

INTERLEUKIN-8 – A MEDIATOR OF INFLAMMATORY LUNG DISEASE?

Diana Smith, Lisa Burrows, and John Westwick

Department of Pharmacology, Hunterian Institute
Royal College of Surgeons
Lincolns Inn Fields, London, WC2A 3PN, U.K.

INTRODUCTION

Chronic inflammation is a feature of the airways of patients exhibiting lung diseases such as asthma, sarcoidosis and bronchitis. This has been recognised from autopsy studies of asthmatic airways (1) and has also been shown to be evident in the airways of asthmatics who have mild asthma (2) or are asymptomatic (3). Bronchial airway hyperreactivity (BHR) to a wide range of pharmacological and physical agents is one of the most characteristic features of asthmatic airways. It is known that intensification of BHR follows antigen challenge in atopic asthmatics (4) and it has been widely presumed that inflammatory events during late-onset reactions may determine the changes in airway reactivity (5).

EOSINOPHILS IN ASTHMA

Eosinophil infiltration is a distinctive feature of the inflammation in asthmatic airways (1) and suggests that the eosinophil may play a role in its pathogenesis (6). There has been shown to be a direct correlation between the degree of blood and tissue eosinophilia in asthma and the severity of the disease (7). Upon antigen challenge in atopic asthmatics, an influx of eosinophils into the airway lumen has been observed in late-phase responders (8) and also an increase in the peripheral blood eosinophil numbers (9). There has also been proposed to be a direct relationship between peripheral and airway eosinophil counts and BHR (10). Therefore, the presence of the eosinophil following allergic asthmatic reactions has been demonstrated, but are they acting as beneficial or harmful cells? Eosinophils contain histaminase, arylsulfatase B, phospholipase D and lysophospholipase which can inactivate histamine, leukotriene C_4, leukotriene D_4, and platelet activating factor (PAF) released from other inflammatory cells. These mediators have all been implicated in the literature as putative mediators in asthma (11); therefore the eosinophil in this capacity would be acting as a beneficial cell. However, eosinophils are also a rich source of lipid mediators including PAF and of toxic proteins such as major basic protein (MBP), eosinophil cationic protein and eosinophil derived neurotoxin (12). Studies using MBP have indicated that bronchial, epithelial and skin cells exposed to MBP in vitro developed a dose-dependent degree of disruption (13). Guinea-pig respiratory epithelium exposed to MBP in vitro at concentrations observed in the sputum of asthmatics (14),

Chemotactic Cytokines, Edited by J. Westwick *et al.*
Plenum Press, New York, 1991

caused epithelial damage with the detachment of ciliated cells (15) and also an increase of airway reactivity to histamine and acetylcholine (16). Epithelial damage and loss of ciliary activity are also prominent features observed in asthma (1).

MEDIATORS AND ASTHMA

Many inflammatory mediators have been implicated in the pathogenesis of asthma such as prostanoids, leukotrienes, bradykinin and PAF (11). PAF is the most plausible mediator which may play a role as it can induce both eosinophil infiltration and BHR both in man and experimental animals (17–20). However, in experimental animals such as the guinea-pig, the induction of maximal eosinophil infiltration into the airways is observed 48 hours after administration of PAF (19); therefore, it is likely that PAF is acting via the induction or generation of another stimulus. Eosinophil production is known to be regulated by T-cells (21) and more specifically by cytokines such as IL-5, GM-CSF and IL-3 (22), so it is possible that PAF is exerting its effect on eosinophils via these cytokines. Another cytokine, namely IL-8 which has been demonstrated to be released during inflammatory reactions and causes activation and chemotaxis of neutrophils (23) may also play a role in the inflammatory cell recruitment seen in the lungs of asthmatics. In addition, IL-8 has also been shown to be chemotactic for human lymphocytes in vitro (24) and when injected intradermally into rat skin, a lymphocyte rich accumulation is observed (25). Lymphocyte accumulation occurs in the airways of atopic asthmatics following allergen challenge (26). As IL-8 is expressed from stimulated alveolar macrophages (27) and epithelial cells (28), local generation of IL-8 in the lungs could contribute to accumulation of inflammatory cells within the airways.

We have examined the possibility that IL-8 may induce inflammatory cell accumulation and induce bronchial hyper-reactivity in an experimental model of asthma.

EXPERIMENTAL STUDIES

a. *Intraperitoneal injection of IL-8*

Administration to guinea-pigs of IL-8 (10 μg) by the intraperitoneal route induced an infiltration of T lymphocytes in the airways as assessed by broncho-alveolar lavage. The peak lymphocyte migration observed in the bronchoalveolar lavage fluid (BALF) was at 4 hours (Figure 1a), IL-8 (10 μg) treated animals having lymphocyte numbers of 0.7±0.2 compared to control animals, 0.1±0.02 ($P<0.001$). At this time point there was no significant increase in the total cell numbers obtained. The lymphocyte migration preceded an infiltration of eosinophils in IL-8 treated animals which was seen to be maximal at 24 hours; $4.4\pm0.7 \times 10^6$ eosinophils compared to $0.6\pm0.4 \times 10^6$ eosinophils in vehicle-treated animals (Figure 1b). The increase in eosinophil numbers was paralleled by a two-fold increase in total cell counts. This effect of IL-8 was dose-related between the doses of 1-30 μg (Figure 2a).

Eosinophil infiltration into the airways was also assessed by histology. Paraffin embedded lung lobes were sectioned transversely. A dose-related (3-30μg/animal) increase of eosinophils surrounding the bronchi was observed (Figure 2b). This increase in eosinophils surrounding the bronchi of IL-8 treated animals was significant at 4 hours (IL-8, $17.95\pm4.4 \times 10^6$; BSA, $10.2\pm1.17 \times 10^6$) and 24 hours post injection when compared to control animals (IL-8, $22.4\pm3.3 \times 10^6$; BSA, $8.4\pm0.6 \times 10^6$). It is interesting to note that no increase in eosinophil numbers

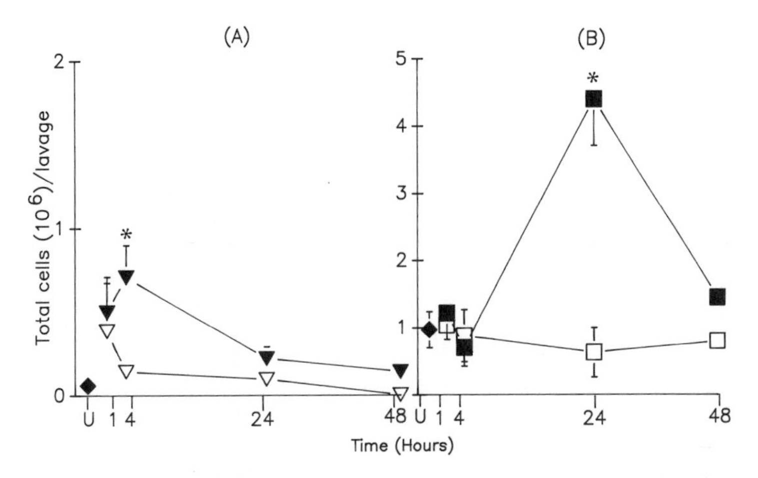

Figure 1: Panel A represents the time course of lymphocyte infiltration observed in BALF following intraperitoneal injection of 10µg of IL-8 (▼) or vehicle control, 0.25% BSA (▽) in the guinea-pig (n=3-8), (◆) untreated (n=4). Panel B represents the time course of eosinophil infiltration observed in BALF following intraperitoneal injection of 10µg IL-8 (■) or vehicle control, 0.25% BSA (□) in the guinea-pig, (◆) untreated (n=4). Stars represent a statistical significance of at least P<0.001.

in the alveoli was observed in IL-8 treated guinea-pigs (Figure 3a). The lung function of these animals which received IL-8 (30 µg) was assessed by an automated pulmonary monitoring system (PMS, Mumed Ltd, UK). This was to determine whether a direct correlation between eosinophil infiltration and changes in airway reactivity could be seen, as has been proposed in humans (10). At the time point when the eosinophil infiltration was maximal in the BALF (24 hours),

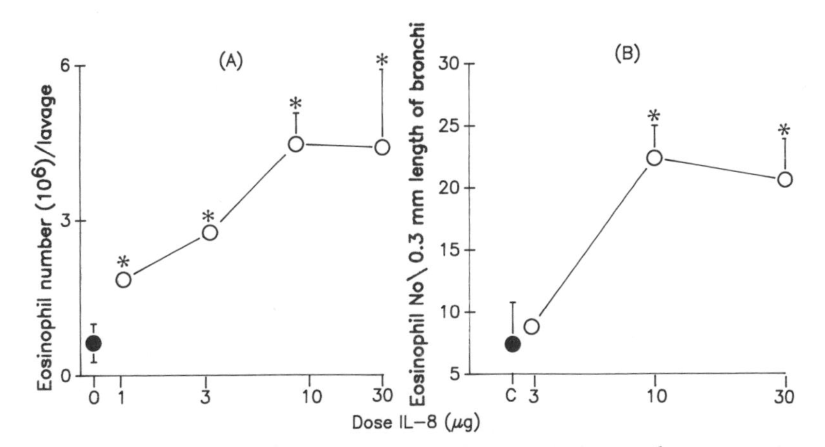

Figure 2: Panel A represents the total eosinophil number (x 10^6) observed in BALF 24 hours following intraperitoneal injection of BSA vehicle (●) and increasing doses of IL-8 (○) in the guinea-pig (n=3-8).
Panel B represents the eosinophil numbers counted per 0.3 mm length of bronchi in paraffin embedded lung lobes 24 hours following intraperitoneal injection of BSA vehicle (●) and increasing doses of IL-8 (○) in the guinea-pig (n=3-8). * represents a statistical significance of at least P<0.05.

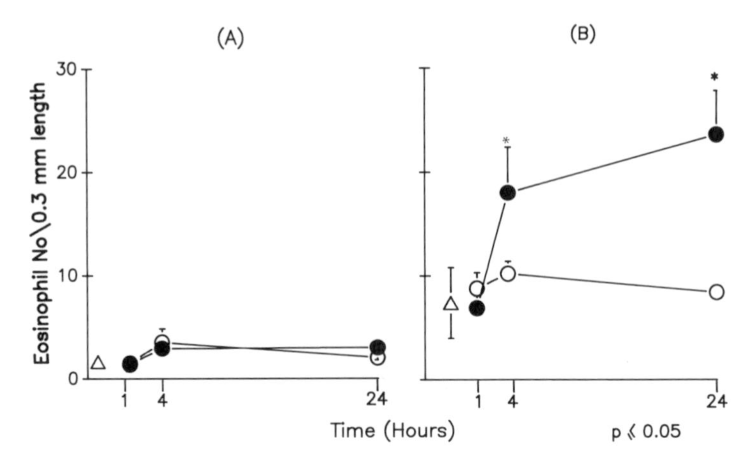

Figure 3: Eosinophil numbers counted per 0.3mm length of alveoli (panel A) and bronchi (panel B) in paraffin embedded lung lobes 24 hours following intraperitoneal injection of BSA vehicle (○) or IL-8 (●) in the guinea-pig (n=3-6). The open triangles depict eosinophils in untreated animals (n=7). * represents a statistical significance of at least $P<0.05$.

airway resistance and dynamic compliance was calculated in anaesthetized animals prepared for the measurement of lung function. There was no change in airway reactivity to an intravenous injection of the airway spasmogen histamine (1–3.2 μg/kg) between a group of animals which had received IL-8 and the vehicle control (0.25% BSA). Therefore, although significant airway eosinophil infiltration was observed both in the BALF and also by histology, there was no correlation with an increased airway reactivity to the spasmogen histamine.

b. *Inhalation of IL-8 via aerosol*

The effect of IL-8 upon eosinophil infiltration into the airways when administered by the aerosol route was also investigated. Guinea-pigs were placed in perspex chambers and exposed over a 30 minute period to an aerosol of either IL-8 (50 μg) or vehicle (0.25% BSA). The aerosol was generated in a deVilbiss nebuliser by passing compressed air at a rate of 7 l/min over a volume of 5 ml of solution. Approximately 0.01% of the total solution has been calculated to reach the guinea-pig airways (29). At various time points after exposure (0, 4, 24 and 48h), airway reactivity to intravenous histamine was measured and the lungs were subsequently lavaged. At 4 hours after exposure, the lymphocytes were elevated in both the vehicle and IL-8 exposed animals. There was no change in the eosinophil numbers when compared to a group of naive animals (time 0 hours). At 24 hours, the lymphocyte numbers had declined to those observed in naive animals. However, both the total eosinophil numbers and the percentage of eosinophils in the IL-8 treated animals had significantly increased when compared to the BSA vehicle animals; thus, in BSA treated animals (n=4), the total eosinophil number was $1 \pm 0.2 \times 10^6$ compared to $6.7 \pm 0.7 \times 10^6$ in IL-8 treated animals (n=6, $P<0.001$, Figure 4) and 12.6±2.9% eosinophils in BSA treated animals compared to 25±2.9% in IL-8 treated animals ($P<0.01$). When the airways were lavaged at 48 hours, there was an even greater increase in both the total number of eosinophils and the percentage eosinophils in the IL-8 treated group when compared to the BSA vehicle exposed animals; thus, the total eosinophil number in BSA treated animals was $2.8 \pm 0.7 \times 10^6$ (n=5) compared to

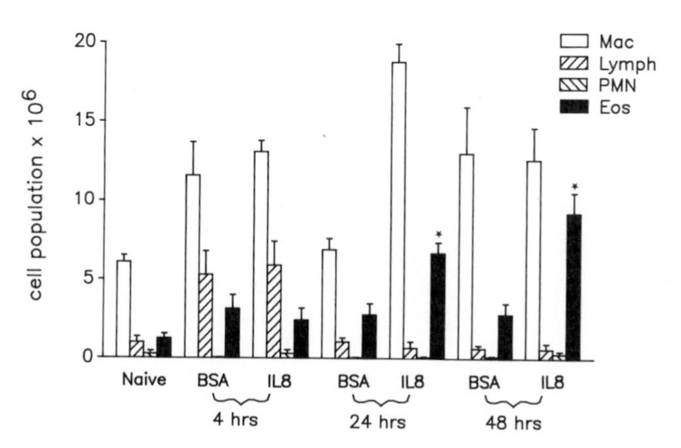

Figure 4: Total cell numbers (x 10^6) observed in BALF following aerosol exposure to BSA vehicle or IL-8 in the guinea-pig. Macrophages (open bars), lymphocytes (right cross hatch bars), neutrophils (left hatch bars) and eosinophil numbers (solid bars) for naive animals and for groups of animals 4, 24 & 48 hours post challenge are depicted (n=6). * represents a statistical significance of at least P<0.05.

9.2±1.3 x 10^6 in IL-8 treated animals (n=6, P<0.005), and 16.5±0.7% eosinophils compared to 40.2±4.1% (P<0.002).

There was no change in the number or percentage of both alveolar macrophages and neutrophils at any time point in both the vehicle treated BSA group and the IL-8 treated group. Therefore, aerosol exposure of guinea-pigs to IL-8 induced a selective increase in eosinophil numbers. Surprisingly, there was no change in neutrophil numbers at any time point, even though neutrophils express receptors for IL-8 and they are highly chemotactic in response to IL-8 in vitro and when IL-8 is injected intradermally into the skin of guinea-pigs. On the other hand, eosinophils are not chemotactic or activated in response to IL-8 in vitro, although they express surface IL-8 receptors. Therefore, it is likely that the delayed onset eosinophil accumulation observed in the airways of guinea-pigs is via the activation of another cell type, and possibly the generation of another mediator which is responsible for the accumulation of the eosinophils.

When the lung function was assessed in these groups of animals, at all time points (4, 24 & 48 hours) there was no significant increase in the airway reactivity to an intravenous injection of histamine (1-3.2 μg/kg). So, both intraperitoneal injection and aerosol exposure to IL-8 induced an eosinophil infiltration into the airways of guinea-pigs without causing any change in the lung reactivity. This is similar to results obtained after PAF challenge or antigen challenge by aerosol in the guinea-pig where a selective increase in eosinophils is observed without the increase in airway reactivity (19,30).

PLATELETS AND ASTHMA

It has been suggested that platelets may be involved in respiratory disease (31-33) and that bronchoconstrictor compounds released from platelets may be important in the pathogenesis of asthma. Evidence for platelet involvement in asthma includes the detection of platelet-derived products such as platelet factor 4, beta thromboglobulin, or thrombospondin in the plasma of allergic subjects

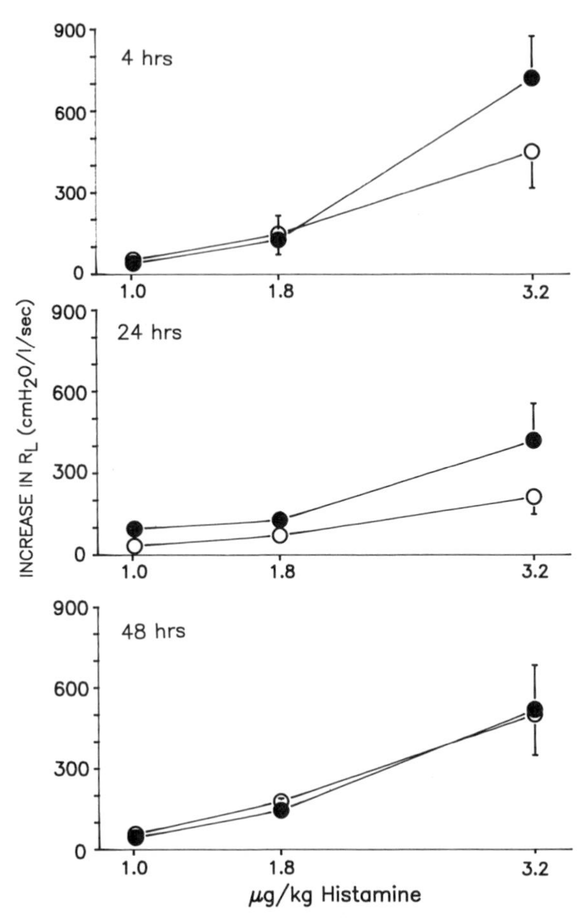

Figure 5: Dose related increase of airway resistance (R_L) in response to an intravenous injection of histamine (1 – 3.2 μg/kg) in animals exposed to either an aerosol of BSA vehicle (0.25%) (○) or IL-8 (50μg) (●) at 4, 24 or 48 hours post challenge.

undergoing allergen challenge (34,35). Recent data indicates that the platelet is a very rich source of an IL-8 related peptide known as CTAPIII which is clipped by monocyte derived protease to neutrophil activating peptide II (36). Although this peptide is about 100 fold less potent than IL-8 as an activator of neutrophils, its precursor CTAPIII is present in α granules at a high concentration of approximately 15 μg/ml of blood. Therefore platelets have the potential of being a very rich source of IL-8 like activity which is not dependent upon protein synthesis and thus explosive generation of IL-8 is possible.

Acknowledgements: We are grateful to the Wellcome Trust and the MRC for financial support, and to Dr Ivan Lindley for recombinant IL-8.

REFERENCES

1. Dunnill, M.S. 1960. The pathology of asthma, with special reference to changes in the bronchial mucosa. *J. Clin. Pathol.* **13**: 27-33.

2. Laitinen, L.A., M. Heino, A. Laitinen, T. Kava, and T. Haahtela. 1985. Damage of the airway epithelium and bronchial reactivity in patients with asthma. *Am. Rev. Respir. Dis.* **131**: 599-606.
3. Beasley, R., W.R. Roche, G.A. Roberts, and S.T. Holgate. 1989. Cellular events in the bronchi in mild asthma and after bronchial provocation. *Am. Rev. Respir. Dis.* **139**: 806-817.
4. Cockcroft, D.W., R.E. Ruffin, J. Dolovich, and F.E. Hargreave. 1977. Allergen induced increase in non-allergic bronchial reactivity. *Clin. Allergy* **7**: 503-513.
5. Chung, K.F. 1986. Role of inflammation in the hyperreactivity of the airways in asthma. *Thorax* **41**: 657-662.
6. Frigas, E., and G.J. Gleich. 1986. The eosinophil and the pathophysiology of asthma. *J. Allergy Clin. Immunol.* **77**: 527-537.
7. Horn, B.R., E.D. Robin, J. Theodore, and A. Van Kessel. 1975. Total eosinophil counts in the management of asthma. *New Engl. J. Med.* **292**: 1152-1155.
8. De Monchy, J.G.R., H.F. Kauffman, P. Venge, G.H. Koeter, H.M. Jansen, J. Suiter, and K. de Vries. 1985. Bronchoalveolar eosinophilia during allergen-induced late asthmatic reactions. *Am. Rev. Respir. Dis.* **131**: 373-376.
9. Booij-Noord, H.J., N.G.M. Orie, and K. de Vries. 1971. Immediate and late bronchial obstructive reactions to inhalations of house dust and protective effects of disodium cromoglycate and prednisolone. *J. Allergy Clin. Immunol.* **48**: 344-354.
10. Durham, S.R., and A.B. Kay. 1985. Eosinophils, bronchial hyperreactivity and late phase asthmatic reactions. *Clin. Allergy* **15**: 411-418.
11. Barnes, P.J., K.F. Chung, and C.P. Page. 1988. Inflammatory mediators in asthma. *Pharmacol. Rev* **40**: 49-84.
12. Gleich, G.J., and C.R. Adolphson. 1986. The eosinophilic leukocyte: structure and function. *Adv. Immunol.* **39**: 177-182.
13. Gleich, G.J., E. Frigas, D.A. Loegering, D.L. Wasson, and D. Steinmuller. 1979. Cytotoxic effects of the eosinophil major basic protein. *J. Immunol.* **123**: 2925-2927.
14. Frigas, E., D.A. Loegering, G.O. Solley, G.M. Farrow, and G.J. Gleich. 1981. Elevated levels of eosinophil granule major basic protein in the sputum of patients with bronchial asthma. *Mayo Clinic. Proc.* **56**: 345-353.
15. Flavahan, N.A., N.R. Slifman, G.J. Gleich, and P.M. Vanhoutte. 1988. Human eosinophil major basic protein causes hyperreactivity of respiratory smooth muscle: Role of the epithelium. *Am. Rev. Respir. Dis.* **138**: 685-688.
16. Frigas, E., D.A. Loegering, and G.J. Gleich. 1980. Cytotoxic effects of the guinea-pig eosinophil major basic protein on tracheal epithelium. *Lab. Invest.* **42**: 35-43.
17. Henocq, E., and B.B. Vargaftig. 1988. Skin eosinophilia in atopic patients. *J. Allergy Clin. Immunol.* **81**: 691-695.
18. Cuss, F.M., C.M.S. Dixon, and P.J. Barnes. 1986. Effects of inhaled platelet-activating factor on pulmonary function and bronchial responsiveness in man. *Lancet* **2**: 189-192.
19. Sanjar, S., S. Aoki, K. Boubekeur, I.D. Chapman, D. Smith, M.A. Kings, and J. Morley. 1990. Eosinophil accumulation in pulmonary airways of guinea-pigs induced by exposure to an aerosol of platelet-activating factor: Effect of anti-asthma drugs. *Br. J. Pharmacol.* **99**: 267-272.
20. Coyle, A.J., S.C. Urwin, C.P. Page, C. Touvay, B. Villain, and P. Braquet. 1988. The effect of the selective PAF antagonist BN 52021 on PAF-induced and antigen-induced bronchial hyperreactivity and eosinophil accumulation. *Eur. J. Pharmacol.* **148**: 51-58.
21. Basten, A., and P.B. Beeson. 1970. Mechanism of eosinophilia. II. Role of the lymphocyte. *J. Exp. Med.* **131**: 1288-1294.
22. Vadas, M.A., A.F. Lopez, M.F. Shannon, and I. Clark-Lewis. 1989. Regulation of eosinophilo-poiesis in man. In Eosinophils in Asthma. J. Morley, and I.G. Colditz, editors. Academic Press, London. 1-11.
23. Baggiolini, M., A. Walz, S.L. Kunkel. 1990. Neutrophil activating peptide -1/interleukin 8, a novel cytokine that activates human neutrophils. *J. Clin. Invest.* **84**: 1045-1050.
24. Bacon, K.B., J. Westwick, and R.D.R. Camp. 1989. Potent and specific inhibition of IL-8, IL-1α- and IL-1β-induced in vitro human lymphocyte migration by calcium channel antagonists. *Biochem. Biophys. Res. Commun.* **165**: 349-354.
25. Larsen, C.G., A.O. Anderson, A. Apella, J.J. Oppenheim, and K. Matsushima. 1989. The neutrophil-activating protein (NAP-1) is also chemotactic for T lymphocytes. *Science* **243**: 1464-1466.
26. Corrigan, C.J., A. Hartnell, and A.B. Kay. 1988. T Lymphocyte activation in acute severe asthma *Lancet* i, 1129-1132.

27. Strieter, R.M., S,L. Kunkel, H,J. Showell, and R,M. Marks. 1988. Monokine-induced gene expression of a human endothelial cell derived neutrophil chemotactic factor. *Biochem. Biophys. Res. Commun.* **156**: 1340-1345.
28. Elner, V.M., R.M. Strieter, S.G. Elner, M. Baggiolini, I. Lindley, and S.L. Kunkel. 1990. Neutrophil chemotactic factor (IL-8) gene expression by cytokine-treated retinal pigment epithelial cells. *Am. J. Pathol.* **136**: 4, 745-750.
29. Richerson, H.B. 1972. Acute experimental hypersensitivity pneumonitis in the guinea-pig. *J. Lab. Clin. Med.* **79**: 745-757.
30. Sanjar, S., S. Aoki, A. Kristersson, D. Smith, and J. Morley. 1990. Antigen challenge induces pulmonary airway eosinophil accumulation and airway hyperreactivity in sensitised guinea-pigs: the effect of anti-asthma drugs. *Br. J. Pharmacol.* **99**: 679-686.
31. Maccia, C.A. 1977. Platelet thrombopathy in asthmatic patients with elevated immunoglobulin E. *J. Allergy Clin. Immunol.* **59**: 101-108.
32. Gallagher, J.S., I.L. Bernstein, C.A. Maccia, G.L. Splansky, and H.I. Glueck. 1978. Cyclic platelet dysfunction in IgE-mediated allergy. *J. Allergy Clin. Immunol.* **62**: 229-235.
33. Slater, D.N., J.F. Martin, and E.A. Trowbridge. 1985. The platelet in asthma. *Lancet* i:110-111.
34. Knauer, K.A., L.M. Lichenstein, N.F. Adkinson, and J.E. Fish. 1981. Platelet activation during antigen-induced airway reactions in asthmatic subjects. *New Engl. J. Med.* **304**: 1404-1407.
35. Gresele, P., T. Todisco, F. Merante, and G.G. Nenci. 1982. Platelet activation and allergic asthma. *New Engl. J. Med.* **306**: 549-549.
36. Walz, A., and M. Baggiolini. 1990. Generation of the neutrophil-activating peptide NAP-2 from platelet basic protein or connective tissue-activating peptide III through monocyte proteases. *J. Exp. Med.* 171: 449-454.

SOME ASPECTS OF NAP-1 PATHOPHYSIOLOGY: LUNG DAMAGE CAUSED BY A BLOOD-BORNE CYTOKINE

Antal Rot

Sandoz Forschungsinstitut
Brunner Strasse 59
A-1235 Vienna, Austria

INTRODUCTION

The neutrophil attracting/activating peptide (NAP-1), also called IL-8, was discovered as a result of its chemotactic activity for human neutrophils *in vitro* (1,2,3). In addition, NAP-1, similarly to other known chemotaxins, has potent neutrophil secretory activity *in vitro* (1,2,3). NAP-1 is the first described cytokine which attracts neutrophils and the first known chemoattractant acting on neutrophils, but not monocytes. Therefore, on the basis of its behaviour *in vitro*, one can suggest that NAP-1 and also the recently discovered NAP-2 (4) and NAP-3 (5) (all three belong to the C-X-C cytokine family, exhibit considerable structural homology and have analogous activities *in vitro*) could be responsible for the induction of the inflammatory cellular infiltrates dominated by neutrophils.

Such neutrophil infiltrates are observed in the synovial tissues of the inflamed joints in rheumatoid arthritis, around the heart and muscle lesions in acute myocardial infarction, in lung interstitium in adult respiratory distress syndrome (ARDS), and in such chronic lung diseases as emphysema and idiopathic lung fibrosis, etc. However, the patho-etiological and pathophysiological role of a chemotactic cytokine can hardly be deduced from its activities *in vitro* but rather has to be established on the basis of clinical studies and experimental observations of the molecule's behaviour *in vivo*.

It was shown that NAP-1 can be produced by a wide variety of cells in different locations of the body, in response to several inflammatory mediators (1,3). The most obvious effect of NAP-1 released in the extravascular tissue is the induction of leukocyte emigration into the site of its production. The injection of NAP-1 into an extravascular site such as skin, peritoneal cavity or joint of the experimental animal provides models of localized inflammatory diseases induced by chemotactic cytokines. These *in vivo* models, in contrast to the chemotaxis assay *in vitro*, enable the observation of the response by a mixed and physiologically proportionate leukocyte population affected by factors peculiar to the situation and processes *in vivo* which are required for leukocyte emigration from the blood vessels (6). These include leukocyte margination (influenced mainly by altered parameters of circulation) and leukocyte adherence to the endothelial cells. In addition to the induction of leukocyte emigration, the injected chemotactic cytokines can stimulate the generation of secondary stimuli, including chemoattractants, the effects of which can also be observed in models *in vivo*.

Chemotactic Cytokines, Edited by J. Westwick *et al.*
Plenum Press, New York, 1991

In the process of inflammation and also following the injection of NAP-1, leukocytes marginate and selectively adhere to the endothelial cell lining of the postcapillary venules, small veins and, in some instances, small arteries. Chemoattractants alone usually fail to induce optimal leukocyte margination. The co-injection of NAP-1 with the factors causing vasodilatation, e.g. prostaglandin E_2, results in increased emigration of leukocytes which is probably achieved by the facilitation of their margination (7). The adherence of leukocytes to the endothelial cells is essential for emigration, as leukocyte adherence is a prerequisite for *in vitro* leukocyte chemotaxis. NAP-1 can promote the adherence of leukocytes to the endothelial cells by up-regulating the expression of leukocyte surface CD11/CD18 molecules (leukocyte-dependent endothelial cell adherence). This happens within minutes following NAP-1 stimulation, by rapid mobilization of adhesion molecules from the intracellular pool (8,9). The up-regulated leukocyte adhesion molecules mediate indiscriminate leukocyte adherence to the endothelium in general and therefore cannot cause the selective leukocyte adherence to the endothelium of postcapillary venules involved in inflammation.

Several cytokines, including IL-1 and TNF, as well as LPS, up-regulate the expression *in vitro* of the intercellular adhesion molecule-1 (ICAM-1) and the endothelial leukocyte adhesion molecule 1 (ELAM-1) on the surface of endothelial cells (10,11). The up-regulation of ICAM-1 and ELAM-1 on the surface of endothelial cells adjacent to inflammatory foci is a protein synthesis-dependent process which takes 2 to 4 h (11) and provides the possibility to target leukocytes to the segments of the blood vessel lining affected by inflammation. However, because the neutrophils begin to adhere to the venular endothelial lining as early as 15 minutes following NAP-1 injection, one should postulate the existence of additional factors targeting neutrophil adhesion to the endothelium of affected vessels. Such factors could be derived from mast cells which seem to be the only resident extravascular cells capable of binding NAP-1 and which, upon binding, can release an array of preformed mediators. Only following their adherence to the venular endothelium can the leukocytes respond to the transendothelial chemotactic gradient and emigrate in the extravascular tissues, where they fulfil their role in host defense. Alternatively, the NAP-1 diffusing from the extravascular tissues could stimulate the not yet adherent leukocytes. These stimulated leukocytes remain in circulation and reach the first anatomically determined organ, in most cases the lungs, where they can adhere to the endothelial cells and be a source of tissue damaging substances. In addition, chemotactic cytokines can be released directly into the bloodstream, e.g. NAP-1 by endothelial cells and leukocytes and NAP-2 by platelets, where they can reach considerable concentrations. The systemic effects of the chemotactic cytokine circulating in blood can be examined by its administration i.v. to experimental animals. In this study, recombinant human NAP-1 (rhNAP-1) was administered i.v. to rabbits and the pathological changes in the organs were recorded.

EXPERIMENTAL DESIGN

Chinchilla rabbits (Ivanovas, Kisslegg, Germany) were housed in the animal facility of the Sandoz Forschungsinstitut and treated in accordance with the guidelines for the accommodation and care of animals by the Council of European Communities. Three different NAP-1 administration protocols were used:

A) Single NAP-1 injection

Each of the four experimental rabbits received 100 μg of rhNAP-1 (12) in one i.v. injection. Control rabbits received an i.v. injection of LPS-free PBS which contained less than 1 pg LPS/ml (as tested in the LAL assay). Two treated and two control animals were sacrificed three hours after the injection. The internal organs were studied histologically. Blood samples were taken periodically from the remaining animals; absolute and differential leukocyte counts were performed. The NAP-1 blood levels after the i.v. injection were measured by the ELISA technique and compared with the NAP-1 blood levels after the s.c. injection of 100 μg of NAP-1.

B) Short-term NAP-1 treatment

Rabbits (two animals/experimental group) received at one hour intervals three injections i.v. of 100 μg of rhNAP-1 (total dose 300 μg). Control animals received repeated injections of LPS-free PBS. Three hours after the last injection the animals were sacrificed, and an autopsy and histological examination of the formalin fixed internal organs was performed.

C) Long-term NAP-1 treatment

Rabbits received one daily injection i.v. of rhNAP-1 (40 μg/kg) during 5 consecutive days. Two control animals received LPS-free PBS injections. Two animals were sacrificed three hours following the last injection, another two animals two weeks after the last injection. The internal organs were studied histologically following their formalin fixation.

RESULTS AND DISCUSSION

Single NAP-1 injection

The injection i.v. of NAP-1 caused marked neutrophilia in rabbits (Figure 1). The neutrophil number peaked 1 h after the injection when 70% of all the

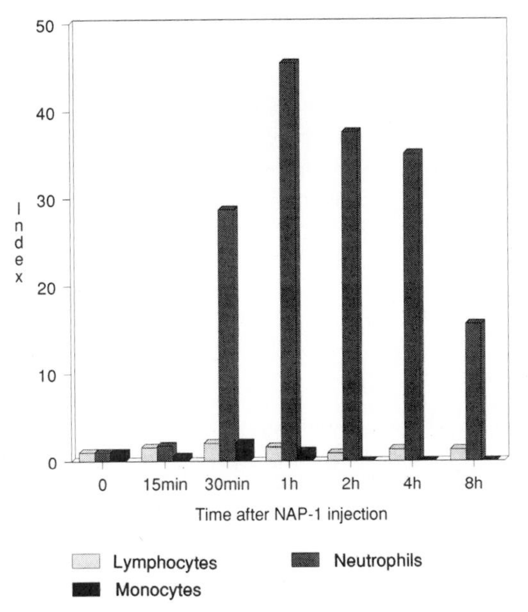

Figure 1. Relative rabbit blood leukocyte counts following NAP-1 injection i.v.

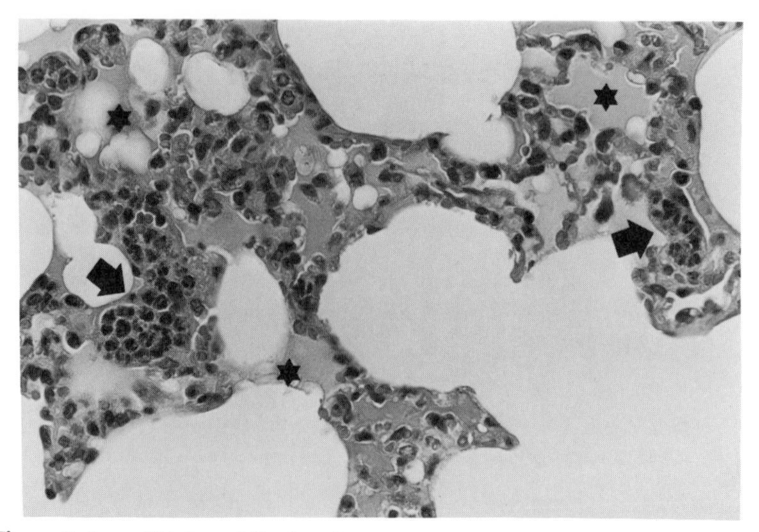

Figure 2. Lung histology following short-term NAP-1 treatment. Arrows indicate intravascular neutrophil agglomerates; stars indicate alveolar edema.

leukocytes in blood were neutrophils. Eight hours after the injection, neutrophil counts remained at a level 15 times above normal. The mechanism by which NAP-1 causes neutrophilia is unknown but is likely to involve the mobilization of the marginal pool and bone marrow reserve. After the i.v. injection of NAP-1, its level in blood initially dropped fast (approximate half-life, 5 minutes); however, after 30 minutes, the rate of the drop progressively declined (half-life, 1-2 h). It is remarkable that 30 minutes after the s.c. injection of NAP-1 its blood levels were almost identical to those seen after the i.v. injection of the same amount of NAP-1. This suggests that NAP-1 produced in extravascular sites can rapidly diffuse into the blood and reach considerable concentrations. The histological examination revealed leuko- and erythrostasis in dilated lung vessels of the NAP-1 treated animals.

Short-term NAP-1 treatment

In spite of the fact that an increased number of neutrophils was seen in the cross-sections of blood vessels of different calibers in different organs of NAP-1 treated animals, only the lungs contained histological features not seen in the control animals. The short-term treatment resulted in the development of intravascular neutrophil aggregates obliterating the lumina of small to medium sized lung blood vessels (Figure 2) and also the appearance of a great number of single-standing neutrophils in lung capillaries. In addition, diffuse interstitial edema, the foci of hemorrhage and intra-alveolar edema (Figure 1), could be seen.

Long-term NAP-1 treatment

The lungs of the animals which received NAP-1 for five consecutive days showed a substantial broadening of the alveolar septa which contained a large number of type II pneumocytes, fibroblasts, and mixed inflammatory cellular infiltrate consisting of small and large mononuclear cells and neutrophils (Figure 3). These changes were seen in all alveoli, however the degree of severity varied. Also, confluent fields of atelectasis and alveoli containing homogenous eosinophilic exudate were observed (Figure 4A). Thus, the short- and long-term NAP-1 treatment resulted in histological changes similar to those seen in the early and late stages of human ARDS, respectively.

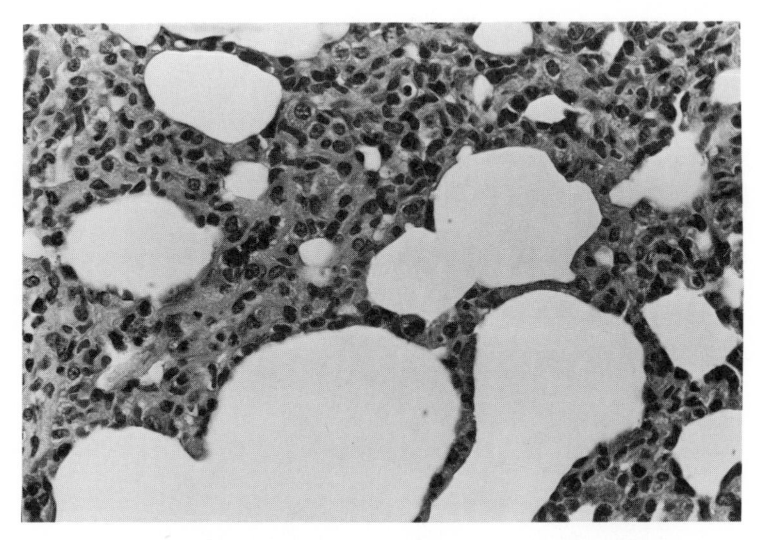

Figure 3. Histological picture of rabbit lungs after long-term NAP-1 treatment. Broad alveolar septa contain mixed leukocyte infiltrate, fibroblasts and type II pneumocytes.

After a recovery period of two weeks, a considerable proportion of rabbit lungs surprisingly returned to their normal histological appearance (Figure 4B). However, the lungs contained several extended areas which consisted of large, distorted alveoli with the air spaces expanded well beyond the normal level; this resembles the histological picture characteristic of human emphysema (Figure 4C). Other signs typical of emphysema including destruction of the alveolar septa and the presence of thin septa, lacking blood vessels could also be observed (Figure 5A). Other lung areas contained alveoli which retained the wide hypercellular septa lined by type II pneumocytes (Figure 4D). The interstitium contained cellular infiltrate consisting of mononuclear cells, neutrophils and fibroblasts, and also fibrous tissue (Figure 5B). The build-up of the collagen fibre, as seen in trichrome stained slides, was especially apparent around the bronchioles and alveolar ducts.

Overall, the i.v. injection of NAP-1 into the experimental rabbits resulted in blood neutrophilia, neutrophil sequestration in lungs and development of a lung pathology which very closely resembles human ARDS and its complications, thus suggesting the involvement of NAP-1 in the pathomechanism of this disease. However, the projection of animal experimental data into the human situation requires great caution since animal models can have specific features which greatly limit any cross-species extrapolation. In addition, the use of a human chemotactic cytokine in animal experimental systems requires a species responsive to the molecule and also the knowledge of the cross-species potency of the cytokine.

Several remarkable examples of species specificity of both cytokines and chemotaxins are known. For example, formylated peptides which are potent chemoattractants for human leukocytes, are not chemotactic for cow, pig, dog, cat, horse and goat neutrophils, whereas leukotriene B_4 (LTB_4) is not chemotactic for rat neutrophils (13,14). Also, human cytokines such as IL-3, IL-4, IL-5, GM-CSF, etc. have limited species cross-reactivity.

In order to establish the cross-species potency of rhNAP-1, we isolated blood neutrophils from chicken, dog, goat, guinea-pig, monkey, mouse, pig, rabbit and rat and studied their chemotactic response to this cytokine *in vitro*. Identical cell separation procedures and conditions of the chemotaxis assay allowed the direct comparison of the chemotactic potency of NAP-1 for the neutrophils of different species. As we reported earlier (15), in spite of the fact that the neutrophils of all

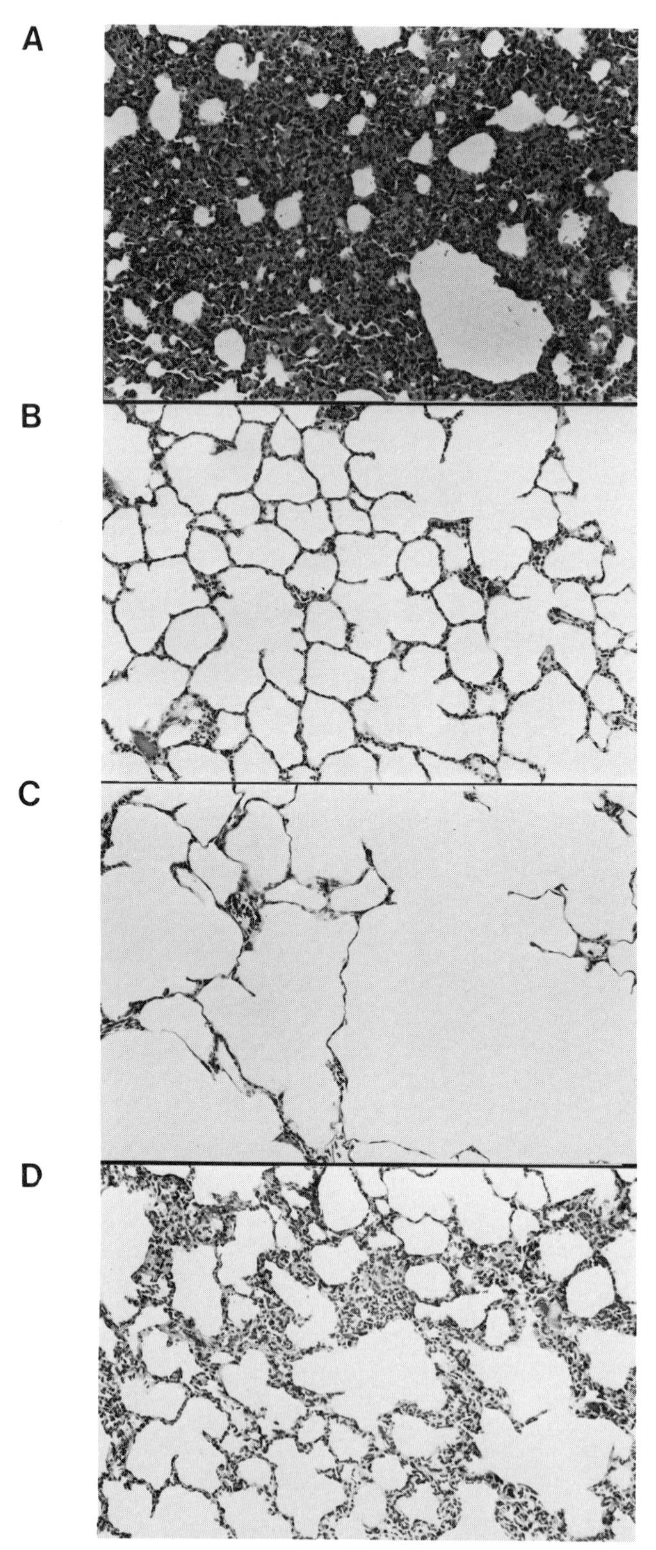

Figure 4. Lung histology after long-term NAP-1 treatment. A: three hours after the termination of treatment; B,C, and D: following a two week recovery period.

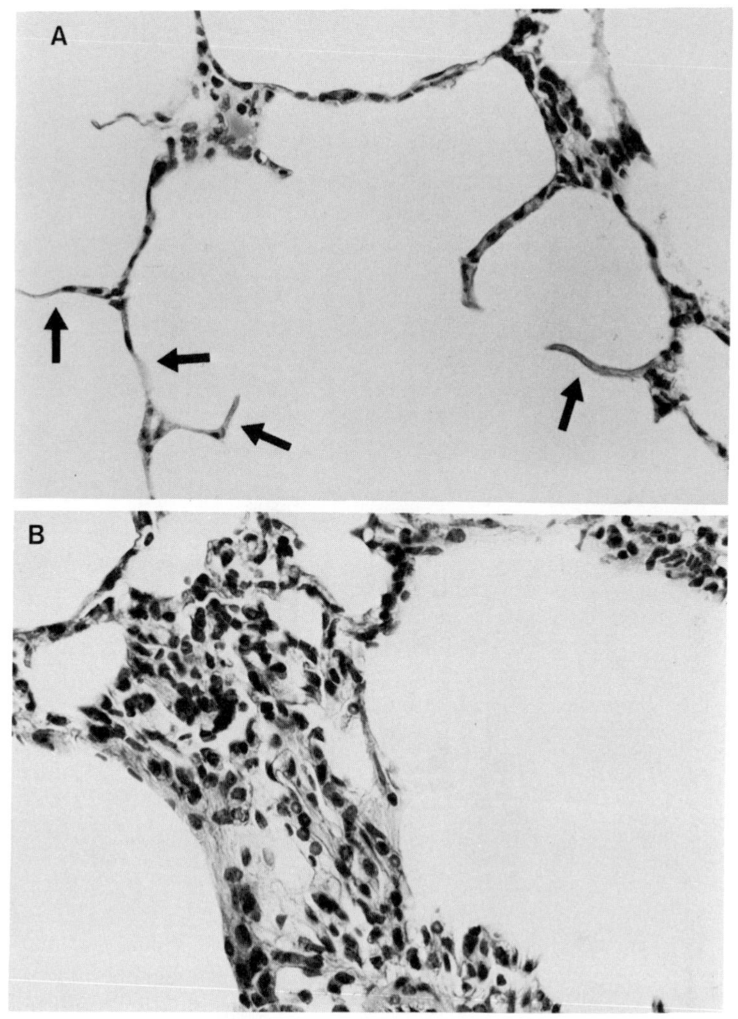

Figure 5. Lung histology after long-term NAP-1 treatment.
A: Histological picture resembling human emphysema. Arrows indicate hypovascular and avascular septa.
B: Wide alveolar septum, containing mixed mononuclear infiltrate, fibroblasts and fibrous tissue.

the tested species migrate to rhNAP-1, the chemotactic potency of NAP-1 for animal neutrophils varies among the species and is considerably less than its potency for human neutrophils. The rhNAP-1 was the least potent for rat and the most potent for monkey neutrophils, whereas rabbit neutrophils were 5 times less sensitive than human cells.

Besides NAP-1, several other molecules capable of directly or indirectly inducing neutrophil sequestration in the lungs were also postulated to play a role in the development of ARDS. These factors include LPS, complement fragments, leukotriene B_4, thromboxane, platelet activating factor, IL-1, TNF, etc. (16). Since NAP-1 is generated upon LPS, IL-1 and TNF stimulation, it could be the common mediator responsible for the lung damage caused by the injection of these factors into the experimental animals (17, 18). Leukocyte chemoattractants, including C5a,

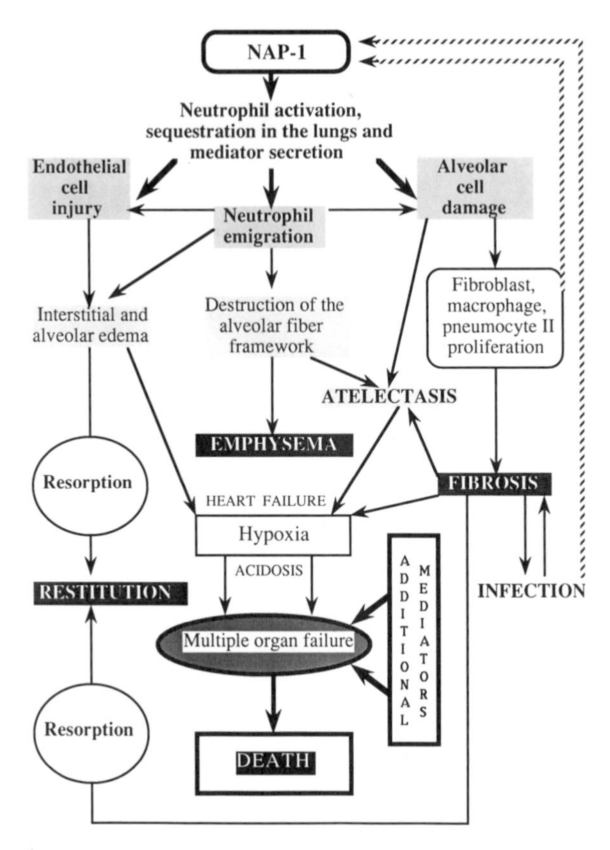

Figure 6. The postulated pathomechanism of lung injury induced by NAP-1.

LTB_4 and PAF, can also induce lung vascular injury under experimental conditions (16, 19). Since C5a can be generated upon activation of complement by LPS, and LTB_4 and PAF are produced by leukocytes after stimulation with a variety of mediators including LPS and chemoattractants, both C5a and LTB_4 can participate in the mediator cascade envisioned in the pathogenesis of ARDS. However, these chemoattractants are short-lived; this raises doubts about their key patho-etiological role in ARDS (16).

It is likely that the ARDS is caused by multiple, "acting in concert" pro-inflammatory factors. The experimental data presented here suggest that the "first violin" of a mediator responsible for neutrophil activation and sequestration in lungs may well belong to NAP-1.

The relevance of this conclusion is supported by the fact that elevated blood levels of NAP-1 were found in serum samples of patients suffering from several diseases known to lead to ARDS (Ceska, M., personal communication).

Thus, the stimulation of neutrophils by NAP-1, their adherence to the lung microvasculature lining, and the secretion of lysosomal enzymes and oxygen radicals trigger a cascade of events which can lead to severe lung injury, ARDS and, in synergism with other factors (such as LPS, TNF, etc.), to multiple organ failure and death (Figure 6).

Acknowledgements: The excellent technical assistance of Ms. Marion Zsak and Mr. Kamillo Thierer, the performance of NAP-1 ELISA by Dr. Miroslav Ceska and fruitful discussions with Drs. H. Walzl, I. Lindley and E. Liehl are gratefully acknowledged.

REFERENCES

1. Baggiolini, M., A. Walz, and S.L. Kunkel. 1989. Neutrophil activating peptide-1/interleukin 8, a novel cytokine that activates neutrophils. *J. Clin. Invest.* **84:** 1045-1049.
2. Matsushima, K. and J.J. Oppenheim. 1989. Interleukin 8 and MCAF: Novel inflammatory cytokines inducible by IL-1 and TNF. *Cytokine.* 1: 2-13.
3. Leonard, E.J. and T. Yoshimura. 1990. Neutrophil attractant/activation protein-1 [NAP-1 (IL-8)]. *Am. J. Resp. Cell Mol. Biol.* 2: 479-486.
4. Walz, A., B. Dewald, V. Tscharner and M. Baggiolini. 1989. Effects of the neutrophil-activating peptide NAP-2, platelet basic protein, connective tissue-activating peptide III, and platelet factor 4 on human neutrophils. *J. Exp. Med.* 170: 1745-1750.
5. Schröder, J-M., N.L.M. Persoon, and E. Christophers. 1990. Lipopolysaccharide-stimulated human monocytes secrete, apart from neutrophil-activating peptide 1/interleukin 8, a second neutrophil-activating protein. *J. Exp. Med.* 171: 1091-1100.
6. Colditz, I.G. 1985. Margination and emigration of leucocytes. *Surv. Synth. Path. Res.* 4: 44-68.
7. Foster A., D.M. Aked, J-M. Schröder, and E. Christophers. 1989. Acute inflammatory effects of a monocyte-derived Neutrophil-activating peptide in rabbit skin. *Immunology.* 67: 181-183.
8. Carveth, H.J., J.F. Bohnsack, T.M. McIntyre, M. Baggiolini, S.M. Prescott, and G.A. Zimmerman. 1989. Neutrophil activating factor (NAF) induces polymorphonuclear leukocyte adherence to endothelial cells and to subendothelial matrix proteins. *Biochem. Biophys. Res. Commun.* 162: 387-393.
9. Detmers, P.A., S.K. Lo, E. Olsen-Egbert, A. Walz, M. Baggiolini, and Z.A. Cohn. 1990. Neutrophil-activating protein 1-interleukin 8 stimulates the binding activity of the leukocyte adhesion receptor CD11b/CD18 on human neutrophils. *J. Exp. Med.* 171: 1155-1162.
10. Dustin, M.L., R. Rothlein, A.K. Bhan, C.A. Dinarello, and T.A. Springer. 1986. Induction by IL-1 and interferon-gamma: tissue distribution, biochemistry, and function of natural adherence molecule (ICAM-1). *J. Immunol.* 137: 245-254.
11. Bevilacqua, M.P., S. Stengelin, M.A. Gimbrone, Jr. and B. Seed. 1989. Endothelial leukocyte adhesion molecule 1: An inducible receptor for neutrophils related to complement regulatory proteins and lectins. *Science.* 243: 1160-1165.
12. Lindley I., H. Aschauer, J-M. Seifert, C. Lam, W. Brunowsky, E. Kownatzky, M. Thelen, P. Peveri, B. Dewald, V. von Tscharner, A. Walz, M. Baggiolini. 1988. Synthesis and expression in Escherichia coli of the gene encoding monocyte-derived neutrophil-activating factor: Biological equivalence between natural and recombinant neutrophil activating factor. *Proc. Natl. Acad. Sci. USA*. 85: 9199-9203.
13. Styrt, B. 1989. Species variation in neutrophil biochemistry and function. *J. Leukocyte Biol.* 46: 63-74.
14. Rot, A. 1990. The role of leukocyte chemotaxis in inflammation. In: *Biochemistry of Inflammation,* eds. Evans, S. and J. Whicher, MTP Press, Lancaster. (In press).
15. Rot, A., C. Lam, and I.J.D. Lindley. 1989. Recombinant human NAP-1 shows differing chemotactic and stimulatory activity on cells from different species. *Cytokine.* 1: 147.
16. Demling, R.H. 1990. Current concepts on the adult respiratory distress syndrome. *Circulatory Shock.* 30: 297-309.
17. Tracey, K.J., B. Beutler, S.F. Lowry, J. Merryweather, S. Wolpe, I.W. Milsark, R.J. Hariri, T.J. Fahey III, A. Zentella, J.D. Albert, G.T. Shires, and A. Cerami. 1986. Shock and tissue injury induced by recombinant human cachectin. *Science.* 234: 470-474.
18. Goldblum, S.E., K. Yoneda, D.A. Cohen, and C.J. McClain. 1988. Provocation of pulmonary vascular endothelial injury in rabbits by human recombinant interleukin-1. *Infection and Immunity.* 56: 2255-2262.
19. Till, G.O. and P.A. Ward. 1986. Systemic complement activation and acute lung injury. *Fed. Proc.* 45: 13-18.

HUMAN MESANGIAL CELL-DERIVED INTERLEUKIN 8 AND INTERLEUKIN 6: MODULATION BY AN INTERLEUKIN 1 RECEPTOR ANTAGONIST

Zarin Brown,[1,2] Lynette Fairbanks,[1,2] Robert M. Strieter,[3]
Guy H. Neild,[2] Steven L. Kunkel,[3] and John Westwick[1]

[1]Department of Pharmacology, Hunterian Institute,
Royal College of Surgeons, London, WC2A 3PN, UK
[2]Department of Renal Medicine, Institute of Urology,
University College and Middlesex School of Medicine, London, UK
[3]Department of Pathology, University of Michigan,
Ann Arbor, Michigan 48109-0602, USA

INTRODUCTION

The position and properties of the mesangial cell confer a major role for these cells in the regulation of renal function (1,2). Mesangial cells reside in the intercapillary space of the glomerulus and are embedded in an extracellular matrix. The interstitial region is unique in that entry of substances from the capillary lumen to the intercapillary space occurs without crossing a capillary basement membrane, the area is therefore well suited for a sieving function.

Glomerulonephritis is characterised by an influx of leucocytes, proliferation of resident mesangial cells and expansion of the mesangial matrix within the glomeruli. The functional consequences of this are impaired filtration, proteinuria and progressive glomerular sclerosis (2). Increased understanding of the role of the mesangial cell in physiology and pathology has been gained principally through the ease with which these cells can now be grown in culture with a high degree of purity (2). In culture, stimulated mesangial cells are able to generate a number of inflammatory mediators including eicosanoids, cytokines, growth factor and bioactive lipids (1) (Figure 1).

We have focused our intention on the cytokines with characteristics that would support a role in either the mesangial cell proliferation or the influx of inflammatory cells that occurs in glomerulonephritis.

Interleukin 6 (IL-6) was first identified as a B cell growth and differentiation factor, but has emerged as a pleiotropic cytokine with multiple biologic activities on different cells (3). Its major biologic activities are induction of acute phase proteins; inhibition of growth and differentiation of a wide variety of tissues and cells, including B and T lymphocytes, myeloma, haematopoietic stem, hepatocytes, fibroblasts and myeloid leukaemia cells (3). A recent report by Suematsu *et al.* established a link between the production of IL-6 and mesangioproliferative glomerulonephritis *in vivo* (4). Transgenic mice carrying the human IL-6 gene fused with a human immunoglobulin heavy chain enhancer developed high

Chemotactic Cytokines, Edited by J. Westwick *et al.*
Plenum Press, New York, 1991

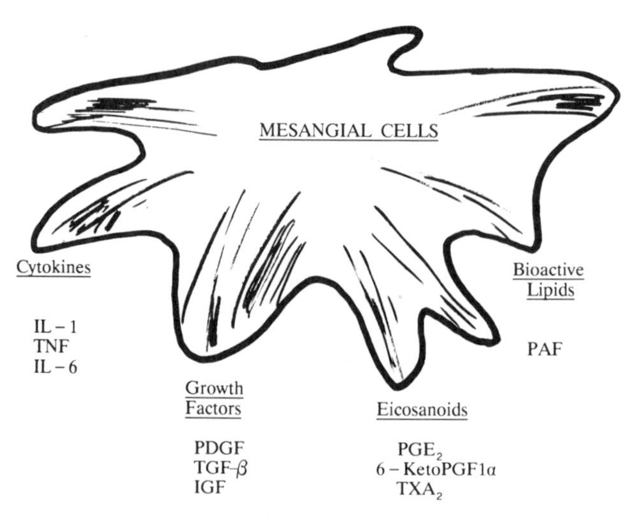

Figure 1. Inflammatory mediators generated by activated mesangial cells.

concentrations of serum IL-6 and massive proteinuria occurred in all mice. Glomeruli from these mice showed marked expansion of mesangium with increase in matrix and mesangial cells. Furthermore, Horii *et al.* have demonstrated that cultured rat mesangial cells when stimulated by LPS are able produce IL-6, and studies from patients with mesangioproliferative glomerulonephritis have shown both an increase in IL-6 in their urine, there was some correlation between levels of urinary IL-6 and disease activity (5). IL-6 is also a potent mitogen for mesangial cell proliferation, *in vitro* (6). The sustained release and action of IL-6 suggest that it is likely to play an important role in glomerulonephritis.

Interleukin 8 (IL-8) was first identified as a potent chemotactic agent and activator of neutrophils, which was initially isolated independently by several groups (7-11) from LPS stimulated monocytes. The purified or recombinant IL-8 was later shown to be a very potent chemotactic agent for human lymphocytes *in vitro* (12,13). Although originally isolated from monocytes, IL-1 or TNF stimulated fibroblasts, endothelial and epithelial cells are a rich source of IL-8 (14,15). Thus these properties, cellular sources, and specificity of induction strongly suggest novel ways in which glomerular inflammation may be initiated and sustained at a local level by IL-8.

MESANGIAL CELL ANATOMY AND PHYSIOLOGY

The mesangium contains two functionally distinct cell types. The predominant cell type phenotypically resembles a smooth muscle cell, the second functionally distinct cell, is a bone marrow-derived cell and constitutes 3-5% of the total cells in the mesangium (1). At the light microscopy level, mesangial cells *in vitro* are stellate shaped cells with an elongated nucleus, irregular cytoplasmic projections and display a prominent microfilament structure. Ultrastructural studies show dense bundles of contractile filaments and dense bodies, similar to those found in smooth muscle cells. The presence of myosin filaments and actomyosin complexes has been confirmed by more recent studies using cultured cell systems (16). The actin bundles span the entire length of the cell processes and the functional consequences of this structural arrangement are that they are capable of contraction. This property of mesangial cells is thought to regulate glomerular filtration rate by altering the surface area available for filtration (1).

PROPERTIES OF THE HUMAN IL-1 RECEPTOR ANTAGONIST

Recently Eisenberg and colleagues have identified IL-1 inhibitory activity in supernatants of human monocytes cultured on adherent IgG or immune complexes (17). This monocyte-derived activity which has been purified and sequenced is a specific IL-1 inhibitor which exists in two forms, a glycosylated form with a molecular weight of 22,000 Da and a non glycosylated form of 17,000 Da. A complementary DNA has been cloned and expressed with the production of recombinant 17,000 mol wt molecules (17). This antagonist blocks IL-1 action by binding to the IL-1 receptor which is present on T-cells and fibroblasts, but does not bind to the IL-1 receptor which is present on B-cells and neutrophils (18).

We have examined the specificity and potency of this novel recombinant peptide (IL-1ra), on the IL-1α and TNFα induced generation of IL-8 and IL-6 by mesangial cells.

METHODS AND RESULTS

Isolation and characterisation of mesangial cells

Portions of macroscopically normal renal cortex were obtained from human kidneys immediately after surgical nephrectomy for renal carcinoma. The isolation of glomerular MC is performed in at least two steps as described by Striker *et al.* (19). Glomerular isolation is the first step of MC culture. This is accomplished by mincing the renal cortex and isolating the glomerular by differential sieving. The second step in MC isolation is the removal of the epithelial cells. This is accomplished by enzymatic dissociation of unencapsulated glomeruli with collagenase. The resultant glomerular fragments were grown on hydrated, collagen gels and maintained in Waymouth's MB752/1 medium supplemented with 15% fetal calf serum. Using this method primary cultures were established as outgrowths from glomeruli by day 8 – 10 and proliferated to confluence by day 15 – 20. Mesangial cell identity was confirmed by using a series of cell markers: immunofluorescence staining of both primary and passaged MC revealed uniform and prominent intracellular longitudinal fibrils for smooth muscle actin (phalloidin-FITC), as well as vimentin, desmin and fibronectin, thus strongly supporting a mesenchymal origin. In addition, MC displayed the characteristic hillock structures in culture as described by Sterzel *et al.* (20). Cells stained with antibodies using the immunoperoxidase method were negative for von Willebrand factor antigen, *Ulex Europaeus* lectin 1 and cytokeratin; thus excluding contamination with endothelial and epithelial cells respectively. Electron microscopy of cell cultures revealed the presence of prominent bundles of peripheral microfilaments with focal attachment bodies (dense bodies) along their course, a feature which is associated with mesangial and smooth muscle cells and is absent in fibroblasts.

Cytokine-induced generation of extracellular IL-8 and IL-6 by mesangial cells

We examined the ability of the pro-inflammatory cytokines IL-1α, TNFα and LPS to induce the gene expression and synthesis of IL-8 and IL-6 from mesangial cells *in vitro*. Growth arrested, confluent MC cultures were stimulated with either hrIL-1α (0.3 – 30 ng/ml), hrTNFα (30 – 1000 ng/ml), LPS (0.1 – 10 μg/ml) or

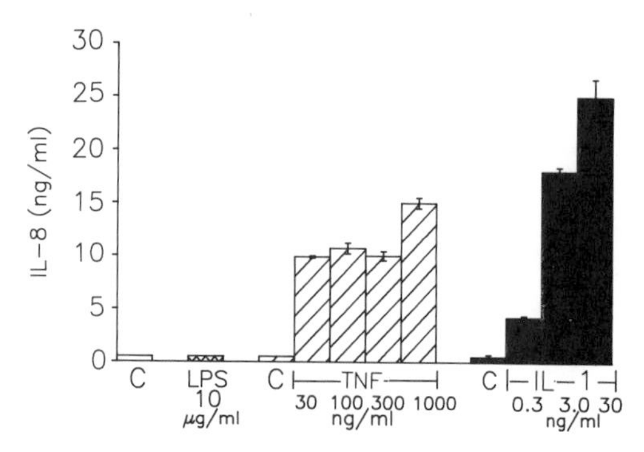

Figure 2. Dose-dependent induction of IL-8 generation by LPS, TNFα and IL-1α treated mesangial cells. The mesangial cells were challenged with the appropriate doses of IL-1α, TNFα, LPS or vehicle control for 24 hours and the supernatants were quantitated for extracellular IL-8 activity by ELISA.

vehicle control for 24 hours. Extracellular concentrations of IL-8 and IL-6 were quantitated by enzyme linked immunoadsorbent assay (ELISA), as previously described by Ceska *et al.* (21) and immunoradiometric assay (IMRA) according to the method of Poole *et al.* (personal communication).

Using these assay systems, we have found that human MC, when stimulated by the appropriate signal, can generate both IL-8 and IL-6. Interleukin 1α, induced a dose-dependent release of extracellular IL-8 and IL-6 by MC's (Figures 2 and 3). Similarly, the addition of TNFα, produced a marked increase in the generation of IL-8 and IL-6, although TNF was about 300 fold less potent than IL-1α (Figures 2 and 3). Interestingly, LPS (10 µg/ml), which has previously been shown to induce the production of IL-8 in a variety of cell types including monocytes, keratinocytes and endothelial cells (14), failed to induce extracellular IL-8 activity in MC.

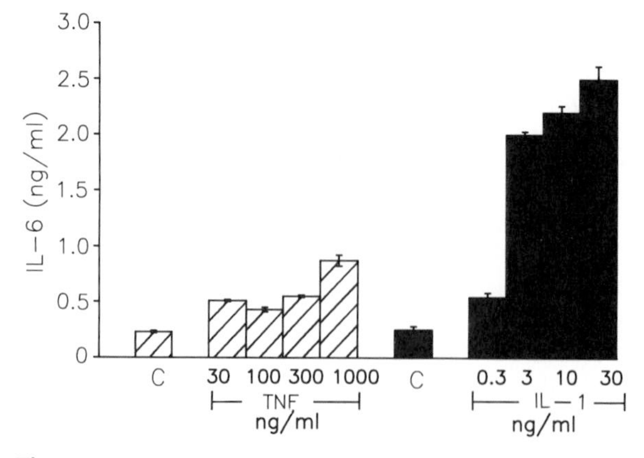

Figure 3. IL-1α and TNF-induced IL-6 production by human mesangial cells. The mesangial cells were challenged with the appropriate doses of IL-1α, TNFα or vehicle control for 24 hours and the supernatants assayed for IL-6 activity by IMRA.

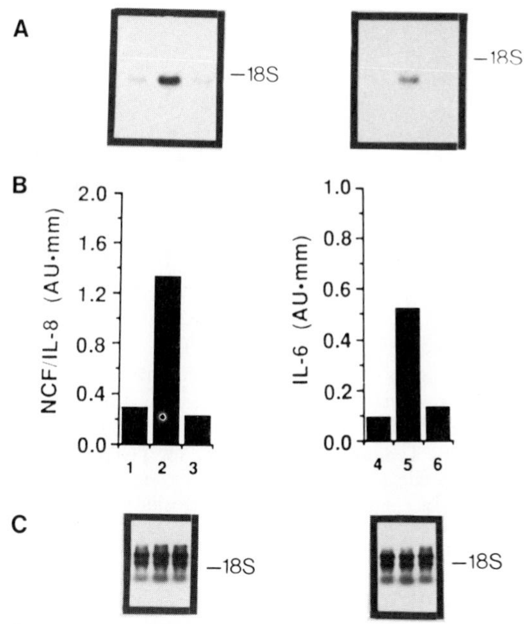

Figure 4. Expression of steady-state IL-8 and IL-6 mRNA by mesangial cells. Human mesangial cells were treated with either IL-1α (30 ng/ml), LPS (10 μg/ml) or vehicle control and total RNA was extracted 6 hours post-challenge (lanes 1–6). Panel (A) shows the Northern blot of the IL-8 (lanes 1–3) and IL-6 (lanes 4–6) mRNA, laser densitometry of each respective Northern blot (panel B) and 28S and 18S ribosomal RNA (panel C) demonstrating equal loading of RNA.

Cytokine-induced expression of IL-8 and IL-6 mRNA

Total cellular RNA from MC was extracted and separated by Northern blot analysis using the method described by Strieter *et al.* (22). Equivalent amounts of total RNA loaded per gel lane were assessed by monitoring 28 and 18S mRNA. Blots were quantitated using laser densitometry. We performed Northern blot analysis to determine the effects of IL-1α on the steady-state IL-8 and IL-6 mRNA expression. Figure 4 shows that IL-1α challenged cells displayed significant steady-state IL-8 and IL-6 mRNA expression. In contrast, neither the non-stimulated nor the LPS (10 μg/ml) treated cells induced detectable IL-8 or IL-6 mRNA expression (Figure 4).

Effects of IL-1ra on cytokine stimulated mesangial cells

Sub-maximal concentration of IL-1α (3 ng/ml) and TNFα (300 ng/ml), selected from dose response studies, were added to confluent growth arrested cultures after a 15 min pre-treatment of the cells with vehicle or IL-1ra (1–1000 ng/ml). Interleukin 1ra inhibited in a dose-dependent manner IL-1α induced extracellular IL-8 and IL-6 generation by MC (Figures 5 and 7). The mean molar ratio for a 50% inhibition for IL-1α induced effects in MC was approximately 15. Calculation of molar ratios of antagonist/agonist demonstrated that IL-1ra was an effective antagonist of IL-1α, but not TNFα induced IL-8 and IL-6 production by MC (Figures 5 and 6).

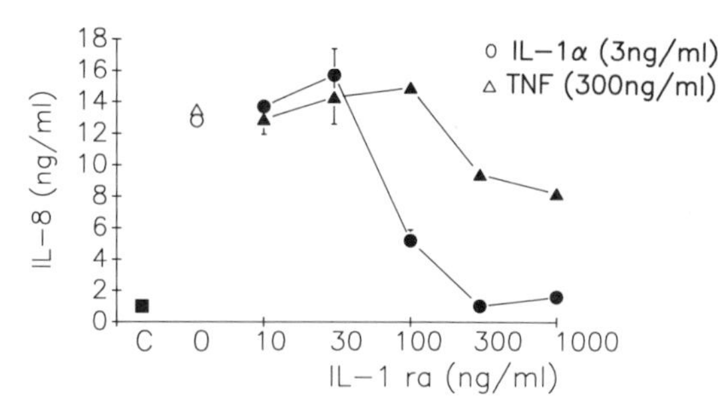

Figure 5. Dose-dependent inhibition by the IL-1 receptor antagonist (IL-1ra) of IL-1α but not TNF induced generation of IL-8 by mesangial cells. Cells were pre-treated with the appropriate dose of IL-1ra for 15 minutes, before the addition of either IL-1α (circles), TNFα (triangles) or vehicle control (square), cultures were incubated for a further 18 hours and the supernatants were assayed for extracellular IL-8 activity.

IL-1ra at concentrations up to 1000 ng/ml exhibited no agonist effects. These results indicate that the recombinant IL-1ra is effective at suppressing IL-1α induced responses in human mesangial cells.

DISCUSSION

We have demonstrated that human mesangial cells *in vitro* produce both IL-6, the mesangial cell mitogen and IL-8, the neutrophil chemoattractant when stimulated with the pro-inflammatory cytokines IL-1α and TNFα.

It has become apparent that the resident glomerular mesangial cell can be an active participant in the inflammatory process. The regulation of cellular interaction via released cytokines is a key event in the initiation and propagation of inflammation. Interleukin 1 and TNF, are the two major cytokines associated with a pro-inflammatory state and both cytokines have been detected in the supernatants of mesangial cells in culture (23,24).

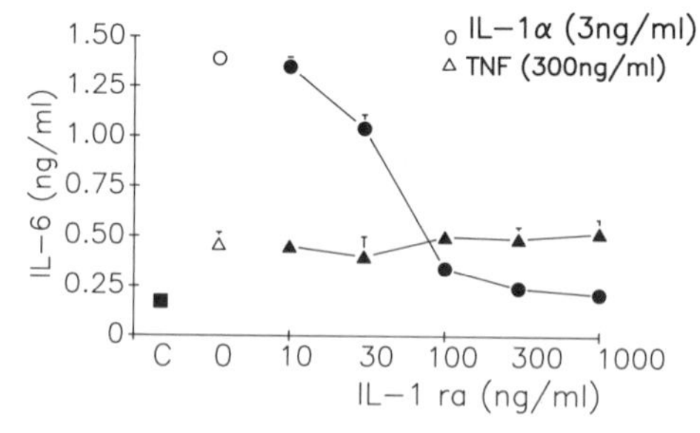

Figure 6. The effect of IL-1 receptor antagonist (IL-1ra) on IL-1α and TNF induced generation of IL-6 by mesangial cells. Cell cultures were pre-treated with the appropriate dose of IL-1ra for 15 minutes before the addition of either IL-1α (circles), TNF (triangles) or vehicle control (square), cultures were incubated for a further 18 hours and the supernatants were quantitated for IL-6 activity.

Although neutrophil accumulation following a delay has been reported at the site of intradermal injection of recombinant IL-1α and TNFα in mice (25) and in rabbits (26), *in vitro* studies have revealed that neither IL-1 nor TNF are chemotactic for human polymorphonuclear cells (15). To provide an explanation for these contradicting observations we and others have demonstrated that the neutrophil stimulating activities of IL-1 and TNF are probably due to the generation of other secreted chemotactic peptides, such as IL-8 (14).

Human mesangial cells stimulated with either IL-1α or TNFα but not LPS demonstrated a kinetic and dose-dependent increase in both IL-8 and IL-6 mRNA expression which paralleled the production of extracellular IL-8 and IL-6. The identification of mesangial cell-derived IL-8 and IL-6, further emphasises that these centrally located, tissue based cells are not passive targets for injury, but active participants in glomerular inflammation.

The finding of mesangial cell-derived IL-8 further establishes that these non-immune cells may play an important role in leucocyte influx to sites of injury. In order for neutrophils to arrive and accumulate in an area of inflammation, they must first marginate and adhere to the endothelium and then migrate to the extravascular space via a chemotactic gradient. This study demonstrates that the mesangial cells is uniquely located to serve as an important source of chemotactic activity to initiate and promote the accumulation and activation of neutrophils at a local level by the generation of IL-8.

Our study also shows that cultured human mesangial cells produce IL-6. Several groups have described the growth promoting effects of IL-6 on cultured rat mesangial cells (5,6). In particular, Horii *et al.* have demonstrated that IL-6 enhanced proliferation of rat MC in the presence of 4% FCS. Furthermore, they found that rat MC's did not produce IL-6 or grow in the culture medium containing less than 4% FCS (5). In contrast to Horii et al. findings, recently Ruef et al. reported a positive mitogenic effect of IL-6 on quiescent MC's, in these experiments IL-6 induced MC growth in the presence of 0.5% FCS (6). In both these instances IL-6 levels were detectable 3 days post stimulation. Our data show that human mesangial cells maintained in serum-free conditions 24 hours prior to and during the experiment were able to produce IL-6 in response to both IL-1α and TNFα, extracellular levels were detectable 4 hours post stimulation with either cytokine. The same experiments carried out in the presence of serum resulted in a 10 fold increase in extracellular IL-6 (results not shown). These results demonstrate that quiescent MC's when activated by IL-1α and to a lesser extent by TNFα are capable of generating IL-6 in serum-free conditions. Thus far little is known about the factors that induce IL-6 production by MC. Our description of IL-6 production following stimulation with IL-1α or TNF provides evidence for the possible mechanism involved in the regulation of IL-6 within the inflamed glomerulus.

Currently there is interest in the possible role of IL-1 and other growth factors to mediate the progression of glomerulonephritis. The potent pleiotropic inflammatory effects of IL-1 on host physiology suggest that its action is tightly regulated *in vivo* (17). The data reported in this study demonstrates that the recombinant monocyte-derived IL-1ra is a effective inhibitor of the inflammatory mediators IL-8 and IL-6 derived from IL-1-activated MC, thus supporting the hypothesis that the actions of IL-1 are stringently regulated at the local level. The physiological relevance of this mechanism of IL-1 regulation is supported by the findings of Arend *et al.* which demonstrate that the antagonist is made by the same cells that make IL-1, and that its synthesis is induced *in vitro* by immune complexes (18). It will be interesting to determine the factors that regulate the gene expression and synthesis of endogenously generated IL-1ra within the normal glomerulus and whether there is reduced expression in glomerulonephritis.

These findings suggest that the mesangial cell is uniquely situated to participate as an important source of inflammatory cytokines which could be responsible both for the accumulation and activation of inflammatory cells as well as the mesangial expansion which occurs in glomerulonephritis. Furthermore the results suggest that a potential therapeutic approach in the control of glomerulonephritis is by the use of an interleukin 1 receptor antagonist.

Acknowledgements: We are grateful to Dr Robert Thompson of Synergen for the supply of the IL-1 receptor antagonist and to the Wellcome Foundation for support.

REFERENCES

1. Mene, P., M. S. Simonson, and M. J. Dunn. 1989. Physiology of the mesangial cell. *Physiol. Rev.* 69:1347 - 1424.
2. Sterzel, R.B and D.H. Lovett. 1990. Interactions of inflammatory and glomerular cells in the response to glomerular injury. In *Contemporary Issues in Nephrology.* B.M. Brenner, J.H. Stein, and C.B. Wilson, editors. Churchill Livingstone, 137-173.
3. Le, J. and J. Vilcek. 1989. Interleukin 6: Multifunctional cytokine regulating immune reactions and the acute phase protein response. *Lab. Invest.* 61:588-602.
4. Suematsu, S., T. Matsuda, K. Aozasa, S. Akira, N. Nakano, S. Ohno, J.I. Miyazaki, K.I. Yamamura, T. Hirano, and T. Kishimoto. 1989. IgG1 plasmacytosis in interleukin 6 transgenic mice. *Proc. Natl. Acad. Sci. USA* 86:7547-7551.
5. Horii, Y., A. Muraguchi, M. Iwano, T. Matsuda, T. Hirayama, H. Yamada, Y. Fujii, K. Dohi, H. Ishikawa, Y. Ohmoto, K. Yoshizaki, T. Hirano, and T. Kishimoto. 1989. Involvement of IL-6 in mesangial proliferative glomerulonephritis. *J. Immunol.* 143:3949-3955.
6. Ruef, C., K. Budde, J. Lacey, W. Northemann, M. Bauman, R.B. Sterzel, and D.L. Coleman. 1990. Interleukin 6 is an autocrine growth factor for mesangial cells. *Kidney Int.* 38:249-257.
7. Yoshimura, T., K. Matsushima, S. Tanaka, E.A. Robinson, E. Appella, J.J. Oppenheim, and E.J. Leonard. 1987. Purification of a human monocyte-derived neutrophil chemotactic factor that has peptide sequence similarity to other host defense cytokines. *Proc. Natl. Acad. Sci. USA* 84:9233-9237.
8. Walz, A., P. Peveri, H. Aschauer, and M. Baggiolini. 1987. Purification and amino acid sequencing of NAF, a novel neutrophil-activating factor produced by monocytes. *Biochem. Biophys. Res. Commun.* 149:755-761.
9. Schröder, J.-M., U. Mrowietz, E. Morita, and E. Christophers. 1987. Purification and partial biochemical characterization of a human monocyte-derived, neutrophil activating peptide that lacks interleukin 1 activity. *J. Immunol.* 139:3474-3483.
10. Van Damme, J., J. Van Beeumen, G. Opdenakker, and A. Billiau. 1988. A novel, NH_2-terminal sequence-characterized human monokine possessing neutrophil chemotactic, skin-reactive, and granulocytosis-promoting activity. *J. Exp. Med.* 167:1364-1376.
11. Westwick, J., S.W. Li, and R.D.R. Camp. 1989. Novel neutrophil-stimulating peptides. *Immunol. Today* 10:146-147.
12. Bacon, K.B., J. Westwick, and R.D.R. Camp. 1989. Potent and specific inhibition of IL-8, IL-1α- and IL-1β-induced *in vitro* human lymphocyte migration by calcium channel antagonists. *Biochem. Biophys. Res. Commun.* 165:349-354.
13. Larsen, C.G., A.O. Anderson, A. Apella, J.J. Oppenheim, and K. Matsushima. 1989. The neutrophil-activating protein (NAP-1) is also chemotactic for T lymphocytes. *Science* 243:1464-1466.
14. Baggiolini, M., A. Walz, and S.L. Kunkel. 1990. Neutrophil-activating peptide-1/interleukin 8, a novel cytokine that activates human neutrophils. *J. Clin. Invest.* 84:1045-1050.
15. Watson, M.L., G.P. Lewis, and J. Westwick. 1988. Neutrophil stimulation by recombinant cytokines and a factor from IL-1-treated human synovial cell cultures. *Immunology* 65:567-572.
16. Becker, C.G. 1972. Demonstration of actomyosin in the mesangial cells of the renal glomerulus. *Am. J. Pathol.* 66:97-110.
17. Eisenberg, S.P., R.J. Evans, W.P. Arend, E. Verderber, M.T. Brewer, C.H. Hannum, and R.C. Thompson. 1990. Primary structure and functional expression from complementary DNA of a human interleukin 1 antagonist. *Nature* 343:341-346.
18. Arend, W.P., H.G. Welgus, R.C. Thompson, and S.P. Eisenberg. 1990. Biological properties of recombinant human monocyte derived interleukin 1 receptor antagonist. *J. Clin. Invest.* 85:1694-1697.
19. Striker, G.E. and L.J. Striker. 1985. Glomerular cell culture. *Lab. Invest.* 53:2:122-122.

20. Sterzel, R.B., D.H. Lovett, H.G. Foellmer, M.C. Perfetto, D. Biemesderfer, and M. Kashgarian. 1986. Mesangial cell hillocks: Nodular foci of exaggerated growth of cells and matrix in prolonged culture. *Am. J. Pathol.* 125:130-140.
21. Ceska, M., F. Effenberger, P. Peichi, and E. Pursch. 1989. Purification and characterisation of monoclonal andpolyclonal antibodies to neutrophil activation peptide (NAP-1). The developement of highly sensitive ELISA methods for the determination of NAP-1 and anti-NAP-1 antibodies. *cytokine* 1:136-136.
22. Streiter, R.M., S.L. Kunkel, H.J. Showell, D.G. Remick, S.H. Phan, P.A. Ward, and R.M. Marks. 1989. Endothelial cell gene expression of a neutrophil chemotactic factor by TNFα, LPS and IL-1β. *Science* 243:1467-1469.
23. Lovett, D.H., K. Resch, and D. Gemsa. 1987. Interleukin-1 and the glomerular mesangium. *Am. J. Pathol.* 129:543-551.
24. Baud, L., J.-P. Oudinet, M. Bens, L. Noe, M.-N. Peraldi, E. Rondeau, J. Etienne, and R. Ardaillou. 1989. Production of tumor necrosis factor by rat mesangial cells in response to bacterial lipopolysaccharide. *Kidney Int.* 35:1111-1118.
25. Granstein, R.D., R. Margolis, S.B. Mizel, and D.N. Sauder. 1986. *In vivo* inflammatory activity of epidermal cell-derived thymocyte activating factor and recombinant interleukin 1 in the mouse. *J. Clin. Invest.* 77:1020-1027.
26. Watson, M.L., G.P. Lewis, and J. Westwick. 1989. Increased vascular permeability and polymorphonuclear leukocyte accumulation *in vivo* in response to recombinant cytokines and a factor produced by interleukin 1-treated human synovial cell cultures. *Br. J. Exp. Pathol.* 70:93-101.

NAP-1/IL-8 IN RHEUMATOID ARTHRITIS

Ivan J.D. Lindley, Miroslav Ceska, and Peter Peichl

Sandoz Forschungsinstitut
Brunner Strasse 59
A-1235 Vienna, Austria

INTRODUCTION

Neutrophil activating peptide 1, or interleukin-8 (NAP-1, IL-8), was isolated almost simultaneously by several groups as a neutrophil chemotactic or activating peptide from supernatants of endotoxin stimulated monocytes (1-5). The major form with 72 amino acids, and an approximate molecular weight of 8.4 kD, has since been shown to have activities on a variety of different cell types. It is chemotactic for neutrophils (PMN) and lymphocytes both *in vitro* and *in vivo* (6,2,4,7), and has been shown to stimulate PMN *in vitro* to degranulate and to exhibit respiratory burst (8,9), and thus has the basic attributes of an inflammatory mediator. Further actions which could be of great importance in inflammatory processes are chemotactic attraction of basophils, the triggering of IL-3 primed basophils to release leukotrienes and histamine (10), spasmogenic activity on airway smooth muscle (11), and effects on PMN adherence to endothelial cells (12).

Cellular sources of NAP-1/IL-8 seem to be as diverse as its activities: in addition to endotoxin-stimulated monocytes, the cell-source from which it was originally isolated, a large number of cell-types have subsequently been shown to produce IL-8 peptide or mRNA when stimulated with endotoxin, interleukin-1 (IL-1) or tumour necrosis factor (TNF). These alternative sources include epithelial and endothelial cells (13,14), fibroblasts (15), chondrocytes (16), hepatocytes (17) and keratinocytes (15). This cytokine therefore has the potential to be produced in a large variety of tissues and organs as a result of inflammatory or infectious stimuli, and to have wide-ranging effects.

The peptide is a member of the large family of "small inducible genes", which share strong structural homology and possess a common bridge structure composed of four cysteine residues. These conserved cysteines give rise to two sub-families, one where the first two cysteines in the sequence are adjacent (the C-C family) and the other, to which IL-8 belongs, where an amino acid is present between them (the C-X-C family). Other members of the NAP-1/IL-8 group include NAP-2, a truncated version of β-thromboglobulin (18), melanoma growth stimulating activity (MGSA) (19), which has recently been dubbed NAP-3, and platelet factor 4 (PF-4) (20). These four peptides have broadly similar chemotactic and activating effects on PMN.

In all, more than 16 members of the superfamily have been identified to date, and most are thought to be involved either directly or indirectly in the

Chemotactic Cytokines, Edited by J. Westwick *et al.*
Plenum Press, New York, 1991

inflammatory response. However, precise biological activity cannot be assigned to them all, as many have been identified only by sequence data from cloned cDNA of stimulated cells, and have not yet been expressed and characterised.

The role of these cytokines in health and disease is, therefore, poorly understood at present, although PF-4 has been studied for several years, and many groups are currently working on NAP-1/IL-8 and the more recently discovered monocyte chemotactic peptide MCP-1 or MCAF (21,22), in terms of cell sources and stimuli, and *in vitro* and *in vivo* effects. In addition some, including our own group, are also examining the *in vivo* pathological role of NAP-1/IL-8 by immunohistochemistry and in situ hybridisation of tissue sections, and determination of levels in serum and tissue samples from hospital patients with well-defined inflammatory diseases.

The migration of leukocytes to a site of injury is an event of primary importance in the inflammatory reaction, and is thought to be mediated by soluble chemotactic factors produced locally. These incoming cells, particularly the neutrophils, are a major source of degradative enzymes (23) and can frequently cause unwanted tissue destruction at the inflammatory focus, inducing clinically recognisable pathological states.

Many chemotactic factors attract a number of different cell types, but NAP-1/IL-8 seems to be unique in being an agent chemotactic for PMN but not for monocytes. Thus it may play a principal role in inflammatory conditions where the cellular infiltrate is predominantly neutrophil-rich, including gout, rheumatoid arthritis, adult respiratory distress syndrome (ARDS), emphysema, glomerular nephritis, myocardial infarction, inflammatory bowel disease and asthma.

We have synthesised a gene encoding the 72 amino acid (monocyte) form of NAP-1/IL-8 and expressed the peptide in E. coli (6). Using the purified recombinant material, we have produced polyclonal and monoclonal antibodies, and have used them to establish a solid phase double ligand ELISA (SPDLE) method (24). With this assay, we have studied levels of NAP-1/IL-8 and antibodies directed against it in the synovial fluid of 38 patients with arthritic disease, including rheumatoid arthritis. In a second study, we determined the same parameters in the serum of 53 RA patients. The results of these two studies are presented here.

EXPERIMENTAL DESIGN

In order to investigate a possible correlation between a chosen marker (in this case, NAP-1 or anti-NAP-1 IgG) and a disease state, one requires objective parameters to measure the severity of disease. For these studies, we used three such parameters.

For the first study, where synovial fluid (SF) was assayed, results were compared with the number of cells in the SF, as increasing cell count is associated with worsening disease and less favourable prognosis.

The second study compared levels of circulating NAP-1 and anti-NAP-1 IgG with the number of arthritic joints, characterised by inflammation, pain, stiffness, swelling and impairment.

In both studies, a comparison was made with circulating levels of quantified C-reactive protein (qCRP), frequently used as a measure of disease severity in RA, since high levels are associated with progressive disease, while induction of clinical remission and control of the underlying disease process is associated with prompt normalisation of the CRP level.

Subjects included as RA patients had classical RA as defined by American Rheumatism Association (ARA) criteria (25). Their clinical examinations were performed by a single observer (P.P.).

METHODS

Synovial fluid samples of 10-50 ml were drawn under sterile conditions from the knees of 38 patients with a variety of arthritic conditions, and small blood samples were taken for routine assay of qCRP level. In the second study, blood samples were taken from RA patients and serum prepared.

NAP-1/IL-8 determinations were performed by solid phase double ligand ELISA within a quantitative evaluation range of 0.2 – 25 ng per ml (24). Briefly, wells of Microtiter plates were coated with immunosorbent-purified polyclonal goat-anti NAP-1 antibodies, and the test sera were then added to the plates and incubated. After washing, goat anti-NAP-1 antibody phosphatase conjugate was used to detect bound NAP-1, which was quantified by comparison to a standard.

Circulating anti-NAP-1/IL-8 IgG antibodies were determined by a similar method, where the samples (diluted 1:10) were incubated with plates which had been pre-coated with recombinant human NAP-1/IL-8. The bound IgG auto-antibodies were then detected with goat anti-human IgG antibody, alkaline phosphatase conjugated. As we do not yet have a human anti-NAP-1 standard IgG, absolute concentration of the antibody could not be defined. The measured levels of antibody are thus expressed as optical density units (OD at 405 nm) from the ELISA assay, giving a reproducible relative figure for the concentration. Full details of this method will be published elsewhere.

Lipopolysaccharide concentrations were determined in all samples using a new automated system (26), and qCRP levels were measured by the nephelometric standard method of Whicher et al (27).

RESULTS

Synovial fluid levels of NAP-1/IL-8 and antibody

All synovial fluid samples contained a low LPS concentration of approximately 10 pg per ml.

Table 1 contains an overview of the results obtained with the 38 patients in this study, classified according to disease, and showing the numbers positive for either NAP-1 or anti-NAP-1 antibody.

NAP-1 was detected in significant quantities in the SF of 11 of the 18 RA patients, while patients with other arthritic conditions showed no detectable NAP-1

Table 1. NAP-1 and anti-NAP-1 levels in synovial fluid of patients with arthritic conditions

Disease (no. of cases)	NAP-1 +ve (conc. NAP-1)	a-NAP-1 +ve (antibody levels)
R.A. (18)	11 (1.25-13.8)	5 (0.296-0.563)
Gonarthrosis (11)	0	3 (0.289-0.457)
Salmonella-induced arthritis (3)	0	0
Gout (3)	0	0
SLE (2)	0	0
Crohn's disease (1)	0	0

NAP-1 concentrations given in ng/ml

Anti-NAP-1 IgG levels given as OD_{405} (see text).

in affected joints. In SF from 5 of the RA patients (none of them NAP-1 +ve) and 3 gonarthrosis patients, antibody was detected at fairly low levels.

The data from the RA patients shown in Table 1 was compared with cell counts in the SF, and the circulating qCRP values. The results, in figure 1, show a highly significant correlation between the two sets of parameters, suggesting that NAP-1 concentration in SF is in some way directly connected with severity of RA.

Serum levels of NAP-1/IL-8 and antibody

A description of the 53 RA patients in this study in terms of age, sex, disease stage and treatment is presented in Table 2.

The serum samples from 14 of the subjects (26.4%) contained detectable NAP-1/IL-8 at low levels of 0.27 to 4.6 ng/ml, and samples taken 8 and 20 days later showed no significant changes (data not shown). In all cases, a low LPS concentration of approximately 10 pg/ml was found.

Healthy control volunteers (n=9) had no detectable NAP-1/IL-8 in their serum, and the anti-NAP-1 IgG ELISA levels obtained with their samples defined the background titre range of O.D.= 0.089 – 0.307 for the antibody. These sera had low levels of qCRP between <8 and 13 mg/l.

In 9 RA patients who were either in complete remission or showing large improvement as a result of therapy, similar low background levels of anti-NAP-1 antibody (0.106 – 0.296) were detected.

However, when active disease was present, higher levels of antibody were seen in the serum, and were proportional to disease severity, as measured by qCRP and by the number of involved joints. Figure 2 shows the highly significant relationship between anti-NAP-1 IgG and qCRP (Figure 2a) and anti-NAP-1 IgG and number of joints affected (Figure 2b). The statistical correlations for this data are presented in Table 3, which demonstrates significance at the $p<0.001$ level for these comparisons.

Table 2. Age, sex, disease stage and treatment of patients included in the study

Mean age (range)	54.2	(27-82)
Sex (F/M)	33/20	
Mean disease duration, years (range)	6.5	(0.6-24)
Disease status:		
Steinbrocker I	21	
II	25	
III	6	
IV	1	
RF +ve/-ve	27/26	
Treatment:		
Antimalarial	15	
Gold salt	16	
D-penicillamine	5	
Steroid/ACTH	11	
Immunosuppressive	17	

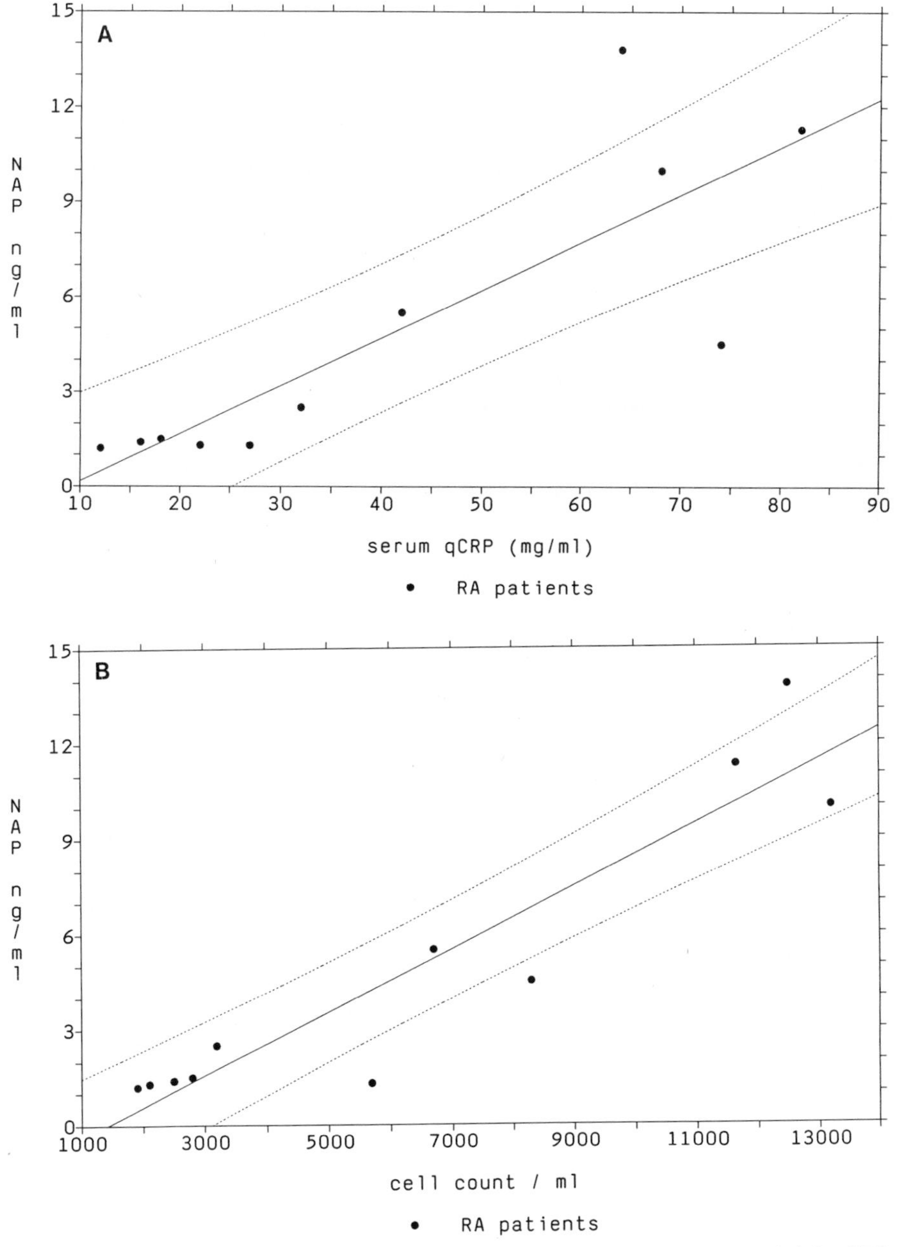

Figure 1. Relationship between NAP-1/IL-8 concentration in synovial fluid and (A) qCRP or (B) number of cells in the joint. Points represent individual patients.

In Figure 3, circulating anti-NAP-1 antibody levels are compared with duration of disease. There was no significant correlation between these two parameters, irrespective of type or duration of treatment. Similarly, there was no significant difference in anti-NAP-1 antibody titres between rheumatoid factor (RF)-positive and RF-negative patients, or male and female patients.

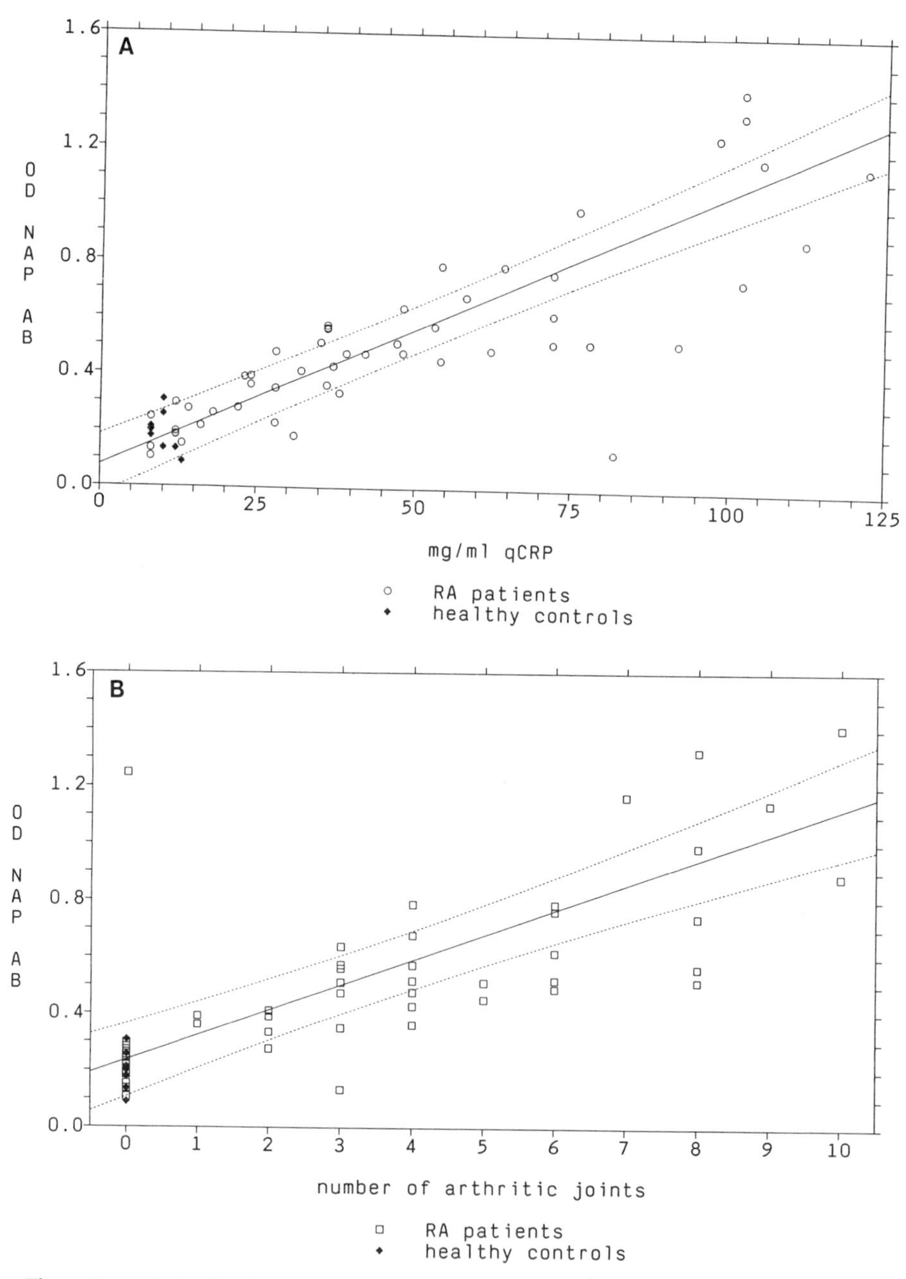

Figure 2. Relationship between circulating anti-NAP-1/IL-8 IgG levels and (A) qCRP or (B) number of arthritic joints. Points represent individual patients. Antibody levels are given as optical density units from the ELISA (see methods).

DISCUSSION

The soluble products of activated macrophages and lymphocytes are thought to play crucial roles in the pathogenesis of RA, as part of, and in combination with, an immune response to an as yet unidentified antigen, and the involvement

Table 3. Statistical correlations between anti-NAP-1/IL-8 (NAP-AB), number of arthritic joints (ART.J.) and qCRP

	qCRP	ARTH.J.	NAP-AB
qCRP	1.000000	0.859965	0.788961
ART.J.	0.859965	1.000000	0.747471
NAP-AB	0.788961*	0.747471*	1.000000

* $p<0.001$

of these inflammatory and chemotactic factors has been well documented. Several of these cytokines have chemotactic activity, and could cause a cellular infiltration into the joint, an event which could in turn enhance inflammatory cytokine levels by production from new cells.

In RA patients, the SF contains large numbers of cells, at levels which reflect the severity of disease in the joint. With worsening disease, the cell number rises and polymorphonuclear leukocytes begin to predominate. This situation is in contrast to that seen in the synovial membrane, where mononuclear cells are the predominant cell types. One of the major activities of NAP-1/IL-8 is chemotactic attraction of PMN, and although other inflammation-associated cytokines like IL-1 and TNF have no such direct activity, they are both known to induce NAP-1 in a variety of different cell types, including those found in the arthritic joint.

The presence of IL-1 (28) and IL-6 (28,29) in synovial fluid has been demonstrated in RA. In spite of correlation between levels of these cytokines and the degree of inflammatory and destructive change occurring, they both lack PMN

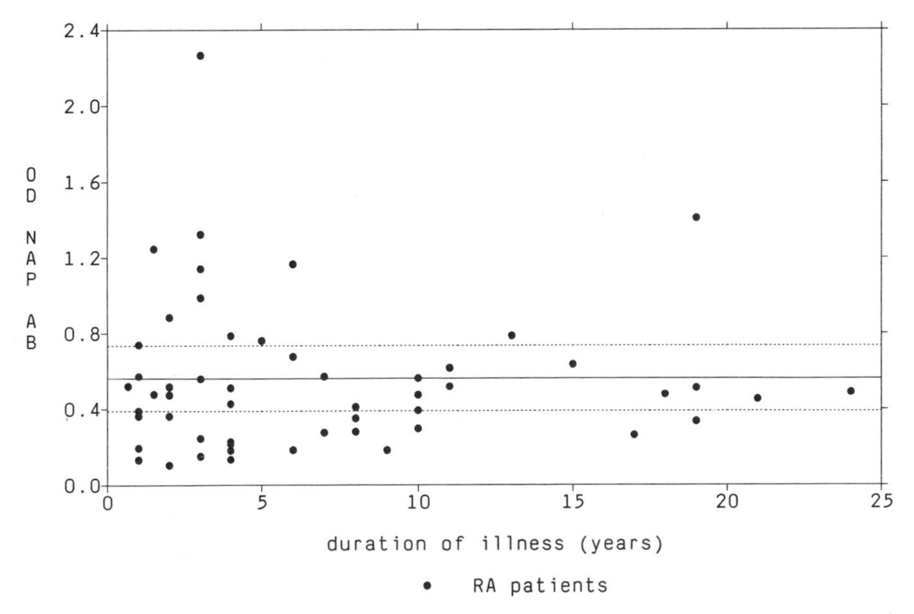

Figure 3. Relationship between circulating anti-NAP-1/IL-8 IgG levels and duration of illness in RA patients. Points represent individual patients. Antibody levels are given as optical density units from the ELISA (see methods).

chemotactic activity and thus another agent is implicated in stimulation of the disease processes of RA.

We have shown NAP-1/IL-8 to be present at significant levels in SF of RA patients, but only at background levels in other diseases where arthritic changes are seen, and the effects of this cytokine may therefore contribute to changes characteristic of RA, including juxta-articular osteoporosis, erosion of subchondrial bone and progressive loss of joint space.

In the RA sera tested, a strong correlation was seen between serological and clinical parameters of disease and the presence of auto-antibodies directed against NAP-1/IL-8, again suggesting a major role for the cytokine in this disease.

Reports of auto-antibodies to cytokines are rare, but anti-IL-1 antibodies have been reported in 17% of RA patients (30), and anti-TNF antibodies have been detected in patients with various arthritic conditions (31). However, these studies were unable to show a direct correlation between antibody titre and disease severity, as we have been able to demonstrate here.

NAP-1/IL-8 could thus be either a stimulator of the inflammatory processes of RA, or a direct result of these processes. However, the progressive joint destruction seen in all cases where high NAP-1 levels were seen, coupled with the PMN-enriched cell count and the highly significant correlation between SF levels of NAP-1 and both serum qCRP and the number of cells in the joint, is strongly suggestive of a more positive, pathogenic role.

The arthritic joint is then probably the major, or only site of high-level NAP-1 production in RA, and it is from here that release of the cytokine occurs into the circulation. The increase in peripheral concentration then triggers production of an autoantibody whose titre parallels the tissue destruction and NAP-1 levels in the joints, and other markers like qCRP. The higher the titre of anti-NAP-1 IgG, the more potent the therapy needed to induce remission or even to stabilise the disease. The highest titres are associated with lack of improvement under systemic steroid treatment, and indicate a need for immunosuppressive therapy. At the present time, it is not clear whether the function of these circulating antibodies is to neutralise possible detrimental effects of free NAP-1 in the periphery, or to stabilise the molecule and preserve some as yet unknown beneficial activity.

More studies are needed to define the role of NAP-1/IL-8 in RA more precisely, and to elucidate its position and effects in the cytokine network. The initial studies presented here demonstrate that NAP-1 has a definite role in the inflammatory processes of the disease, and that the circulating autoantibodies may provide a clinically useful marker for disease severity.

Acknowledgements: We gratefully acknowledge the expert technical assistance of Franz Effenberger, and would like to thank Professor Alain de Weck and Dr. Ekke Liehl for critical discussion of the work.

REFERENCES

1. Walz, A., P. Peveri, H. Aschauer, and M. Baggiolini. 1987. Purification and amino acid sequencing of NAF, a novel neutrophil-activating factor produced by monocytes. *Biochem. Biophys. Res. Comm.* 149: 755-761
2. Schröder, J., U. Mrowietz, E. Morita and E.Christophers 1987. Purification and partial biochemical characterisation of a human monocyte-derived, neutrophil activating peptide that lacks interleukin 1 activity. *J. Immunol.* 139: 3474-3483
3. Yoshimura, T., K. Matsushima, S. Tanaka, E. Robinson, E. Apella, J.J. Oppenheim and E.J. Leonard. 1987. Purification of a human monocyte-derived neutrophil chemotactic factor that shares sequence homology with other host defence cytokines. *Proc. Natl. Acad. Sci. USA* 84: 9233-9237.

4. Van Damme, J., J. Van Beeumen, G. Opdenakker and A. Billeau. 1988. A novel NH_2-terminal sequence- characterised human monokine possessing neutrophil chemotactic, skin reactive, and granulocytosis promoting activity. *J. Exp. Med.* 167: 1364-1376.
5. Kownatzki, E., S. Uhrich and G. Grueninger. 1988. Functional properties of a novel neutrophil chemotactic factor derived from human monocytes. *Immunibiol.* 177: 352-362.
6. Lindley, I., H, Aschauer, J.M. Seifert, C. Lam, W.Brunowsky, E. Kownatzki, M. Thelen, P. Peveri, B. Dewald, V. von Tscharner and M. Baggiolini. 1988. Synthesis and expression in E.coli of a gene encoding monocyte-derived neutrophil-activating factor: Biological equivalence between natural and recombinant neutrophil-activating factor. *Proc. Natl. Acad. Sci. USA* 85: 9199-9203.
7. Larsen, C.G., A.O. Anderson, E. Apella, J.J. Oppenheim and K. Matsushima. 1989. The neutrophil activating protein (NAP-1) is also chemotactic for T lymphocytes. *Science* 243: 1464-1466.
8. Peveri, P., A. Walz, B. Dewald and M. Baggiolini. 1988. A novel neutrophil-activating factor produced by human mononuclear phagocytes. *J. Exp. Med.* 167: 1547-1559.
9. Thelen, M., P. Peveri, P. Kernen, V. von Tscharner, A. Walz and M. Baggiolini. 1988. Mechanism of neutrophil activation by NAF, a novel monocyte-derived peptide agonist. *FASEB. J.* 2: 2702-2706.
10. Dahinden, C.A., J. Kurimoto, A.L. De Weck, I. Lindley, B. Dewald and M. Baggiolini. 1989. The neutrophil-activating peptide NAF/NAP-1 induces histamine and leukotriene release by interleukin 3-primed basophils. *J. Exp. Med.* 170: 1787-1792.
11. Burrows, L.J., P.J. Piper, I. Lindley, M. Baggiolini and J. Westwick. 1990. Characterisation of human recombinant neutrophil activating factor/interleukin 8-induced contraction of airway smooth muscle. In press in: *Molecular and Cellular Biology of Cytokines.* Alan R. Liss, Inc. N.Y. USA
12. Carveth, H.J., J.F. Bohnsack, T.M. McIntyre, M. Baggiolini, S.M. Prescott and G.A. Zimmerman. 1989. Neutrophil activating factor (NAF) induces polymorphonuclear leukocyte adherence to endothelial cells and to sub-endothelial matrix proteins. *Biochem. Biophys. Res. Comm.* 162: 387-393.
13. Elner, V., R. Strieter, S. Elner, M. Baggiolini, I. Lindley and S. Kunkel. 1990. Human retinal pigment epithelial cells produce neutrophil chemotactic factor in response to IL-1 and TNF. *Amer. J. Pathol.* 136: 745-750.
14. Dixit,V.M., S.L. Kunkel, V. Sarma, R.M.Strieter, H.J. Showell, P.A. Ward and R.M. Marks. 1989. Molecular cloning of an endothelial derived neutrophil chemotactic factor: identity with monocyte derived factor. *FASEB. J.* 3: A305
15. Larsen, C.G., A.O. Anderson, J.J.Oppenheim and K. Matsushima. 1989. Production of interleukin-8 by human dermal fibroblasts and keratinocytes in response to interleukin-1 or tumor necrosis factor. *Immunology* 68: 31-36.
16. Van Damme, J., R.A.D. Bunning, R. Conings, R. Graham, G. Russel and G. Opdenakker. 1990. Characterisation of granulocyte chemotactic activity from human cytokine stimulated chondrocytes as interleukin-8. *Cytokine* (in press).
17. Thornton, A.J., R. Strieter, I. Lindley, M. Baggiolini and S.L. Kunkel. 1990. Cytokine-induced gene expression of neutrophil chemotactic factor/interleukin 8 in human hepatocytes. *J. Immunol.* 144: 2609-2613.
18. Walz, A., and M. Baggiolini. 1989. A novel cleavage product of β-thromboglobulin formed in cultures of stimulated mononuclear cells activates human neutrophils. *Biochem. Biophys. Res. Comm.* 159: 969-975.
19. Richmond, A., E. Balentien, H.G. Thomas, G. Flaggs, D.E. Barton, J. Spiess, R. Bordoni, U. Francke and R. Derynck. 1988. Molecular characterisation and chrosomal mapping of melanoma growth stimulatory activity, a growth factor structurally related to β-thromboglobulin. *EMBO J.* 7: 2025-2033.
20. Deuel, T.F., P.S. Keim, M. Farmer and R.L. Heinrikson. 1977. Amino acid sequence of human platelet factor 4. *Proc. Natl. Acad. Sci. USA*. 74: 2256-2258.
21. Furutani, Y., H. Nomura, M. Notake, Y. Oyamada, T. Fukui, M. Yamada, C.G. Larsen, J.J. Oppenheim and K. Matsushima. 1989. Cloning and sequencing of the cDNA for human monocyte chemotactic and activating factor (MCAF) *Biochem. Biophys. Res. Comm.* 159: 249-255.
22. Yoshimura T., N. Yuhki, S.K. Moore, E. Apella, M.I. Lerman and E.J. Leonard. 1989. Human monocyte chemoattractant protein-1 (MCP-1): Full length cDNA cloning, expression in mitogen-stimulated blood mononuclear leukocytes, and sequence similarity to mouse competence gene JE. *FEBS Letts.* 244: 487-493.
23. Baggiolini, M., A. Walz and S. Kunkel. 1989. Neutrophil activating peptide 1/interleukin 8, a novel cytokine that activates neutrophils. *J. Clin. Invest.* 84: 1045-1049.

24. Ceska, M., F. Effenberger, P. Peichl and E. Pursch. 1989. Purification and characterisation of monoclonal and polyclonal antibodies to neutrophil activating peptide: The development of highly sensitive ELISA methods for determination of NAP-1 and anti-NAP-1 antibodies. *Cytokine* 1: 136.
25. Ropes, M.W. 1958. Diagnostic criteria for rheumatoid arthritis. *Ann. Rheum. Dis.* 18: 49-54.
26. Seidl, G. and L. Malacek. 1989. Application of laboratory robotics to the automation of the turbidometric LAL endotoxin assay. *Lab. Robot. and Automation* 1: 215-225.
27. Whicher, J.T., A.M. Bell and P.J. Southall. 1981. Inflammation measurements in clinical management. *Diagn. Med.* 4: 62-80.
28. Bhardwaj, N., U. Santhanam, L.L. Lau, S.B. Tatter, J. Ghrayeb, M. Rivelis, R.M. Steinman, P.B. Sehgal and L.T. May. 1989. IL-6/IFNb2 in synovial effusions of patients with rheumatoid arthritis and other arthritides. *J. Immunol.* 143: 2153-2159.
29. Waage, A, C. Kaufmann, T. Espevik and G. Husby. 1989. Interleukin-6 in synovial fluid from patients with arthritis. *Clin. Immunol. Immunopathol.* 50: 394-398.
30. Suzuki, H., T. Akama, M. Okane, I. Kono, Y. Matsui, K. Yamane and H. Kashiwagi. 1989. Interleukin-1-inhibitory IgG in sera from some patients with rheumatoid arthritis. *Arthritis and Rheumatism* 32: 1528-1538.
31. Fomsgaard, A, M. Svenson and K. Bendtzen. 1989. Auto-antibodies to tumour necrosis factor in healthy humans and patients with inflammatory diseases and Gram-negative bacterial infections. *Scand. J. Immunol.* 30: 219-223.

CLONING OF CAP37, A MONOCYTE-SPECIFIC CHEMOTACTIC FACTOR RELEASED FROM AZUROPHILIC GRANULES OF PMN LEUKOCYTES

JG Morgan,[1] HA Pereira,[2] T Sukiennicki,[1] JK Spitznagel,[2] and JW Larrick[1]. [1]Genelabs Inc, 505 Penobscot Drive, Redwood City, CA 94063, USA. [2]Department of Microbiology & Immunology Emory University School of Medicine, Atlanta GA 30322, USA

Early inflammatory lesions are comprised primarily of polymorphonuclear leukocytes (PMN) followed by infiltration of mononuclear phagocytes. Evidence suggests that one or more factors derived from the PMN may modulate the subsequent chemotaxis of monocytes. In fact an early study described the absolute requirement of PMNs for the subsequent infiltration of mononuclear phagocytes (1).

Previous work from this laboratory demonstrated the antimicrobial activity of a major PMN granule protein of 37 kd [designated Cationic Antimicrobial Protein] (2). Recent work suggests that this protein is a potent mediator of the monocyte specific response (Pereira, A. *et al. J. Clin. Invest.*; in press). This protein is related to the serine class of proteinases and shares significant homology with several other proteinases (e.g. elastase, cathepsin G) derived from the granules of inflammatory leukocytes that are thought to mediate tissue damage during the course of inflammatory diseases. We recently showed that CAP37 lacks proteinase activity, apparently because a mutation (His->Ser) at position 41 destroys the serine proteinase catalytic triad (3). We have isolated a full length cDNA for CAP37 from HL60 cells using probes designed from the N-terminal sequence of CAP37.

1. Page, A.R., and R.A. Good. 1958. A clinical and experimental study of the function of neutrophils in the inflammatory response. *Amer. J. Pathol.* 34: 645.
2. Shafer, W.M., L.E. Martin, J.K. Spitznagel. 1986. Late intraphagosomal hydrogen ion concentration favours the *in vitro* antimicrobial capacity of a 37-kilodalton cationic granule protein of human neutrophil granules. *Infect. Immun.* 53: 651.
3. Pereira, H.C., J.K. Spitznagel, J. Pohl, D. Wilson, J. Morgan, I. Palings, J. W. Larrick. 1990. The 37 KD human neutrophil granule cationic protein shares homology with inflammatory proteinases. *Life Sciences* 46: 189.

THREE HUMAN HOMOLOGS OF MURINE MACROPHAGE INFLAMMATORY PROTEIN-1 (MIP-1)

DR Forsdyke, S Blum, DP Sideris, RE Forsdyke, H Yu, and E Carstens. Departments of Biochemistry, Microbiology, Immunology, Queen's University, Kingston, Ontario, Canada K7L 3N6.

The protein now referred to as MIP-1 was first identified both in mice (by analysis of macrophage supernatants [1]) and in humans (by cDNA cloning; G0S19 [2]). Analysis of the cDNA sequence showed a derived protein (92 amino acids) similar to proteins encoded by what is now known as the "small inducible gene" (SIG) family. Genomic blots showed a complex pattern of hybridization to G0S19 cDNA, indicating at least three genes. These were isolated from genomic libraries and sequenced. The first gene (G0S19-1; Genbank nos: m23178, m32337) corresponds to the cDNA (m23452) which has also been cloned in other laboratories (LD78; AT464) and thus is probably the major G0S19 gene expressed in adult lymphoid cells. The second gene (G0S19-2; Genbank nos: m24110, m32338) encodes a similar derived protein. Removal of the hydrophobic leader sequence (23 amino acids) would leave a 70 amino acid secretory protein with the amino-terminal Ala-Pro dipeptide characteristic of cytokines. Both these genes have patterns of exon-intron organization characteristic of the SIG family. However, the exon 2 open-reading frame of G0S19-3 extends upstream indicating a protein of at least 159 amino acids. The 5' flanks of G0S19-1 and G0S19-2 contain various elements found in cytokine-encoding genes (CK-2/CLE-2), in genes responsive to serum response transcription factor (SRE) and, in the case of G0S19-1, in genes responsible to phorbol ester (Ap1-like elements). G0S19-1 cDNA was inserted into a baculovirus expression vector under the control of the polyhedrin promoter to generate a recombinant (pRP23G0S19-1). Expression was obtained in insect cells and large quantities of an approx. 10kb product were released into the culture medium.

1. Wolpe, S.D. and A. Cerami. 1989. Macrophage inflammatory proteins 1 and 2: members of a novel superfamily of cytokines. *FASEBJ.* 3: 2565-2573.
2. Forsdyke, D.R. 1985. cDNA cloning of mRNAs which increase rapidly in lymphocytes cultured with concanavalin-A and cycloheximide. *Biochem. Biophys. Res. Comm.* 129: 619-625.

SYNTHESIS OF LD78 PROTEIN AND ITS BINDING TO MYELO-MONOCYTIC LEUKEMIA CELL LINES

T Hattori,[1] Y Yamamura,[1] Y Ohmoto,[2] T Nishida,[2] and K Takatsuki[1]
[1]2nd Dept. of Internal Medicine, Kumamoto University Med. School
[2]Research Institute of Otsuka Pharmaceutical Company, Japan

LD78 protein has sequences similar to mouse MIP and related proteins (1). The LD78 gene expressions were found in various hematopoietic cell lines, though the protein synthesis was detected only in HTLV-1 infected cell lines (2). We here describe that the protein synthesis could also be detected in myelo-monocytic leukemia cell lines (HL-60, U937) by stimulation, and radio-labelled LD78 protein bound to U937 cells most efficiently. LD78 protein was synthesized in *Escherichia coli* and the protein was purified by Q-Sepharose FPLC, Heparin FPLC, Octadecyl-4PW/HPLC and SP-5PW/HPLC. Either U937 or HL-60 cells were stimulated with PMA for 2 days. RNA, or both culture supernatants and cellular extracts were obtained from the cells for Northern and Western blot assays. OCT801, specific antisera against LD78 protein, was used to detect the synthesis of LD78 protein by Western blot assay (2). For binding assays, 5μg of LD78 protein was iodinated by use of a Bolton-Hunter method. ^{125}I-LD78 (20 ng, $1x10^5$ cpm) was incubated with $2x10^6$ of various cell lines (U937, MJ, Jurkat). To correct for nonspecific binding, incubations were also performed in the presence of 100-fold of unlabelled protein. Western blot assay showed that unstimulated HL-60 and U937 cells did not synthesize LD78 protein, but did synthesize after the stimulation. The protein (7.8 kD) was detected in both culture supernatants and in cell lysates. ^{125}I-LD78 protein bound to U937 cells most efficiently (2,220 cpm). However, the binding to MJ and Jurkat was 830 and 1,100 cpm respectively. Therefore, we think that LD78 protein is biologically active on myelo-monocytic cells in autocrine manner.

1. Obaru, K., T. Hattori, Y. Yamamura, K. Takatsuki, H. Nomiyama, S. Maeda, and K. Shimada. 1989. A cDNA clone inducible in human tonsillar lymphocytes by a tumor promoter codes for a novel protein of the β-thromboglobulin superfamily. *Mol.Immunol.* 26: 423-426.
2. Yamamura, Y., T. Hattori, K. Obaru, K. Sakai, N. Asou, K. Takatsuki, Y. Ohmoto, H. Nomiyama and K. Shimada. 1989. Synthesis of a novel cytokine and its gene (LD78) expressions in hematopoietic fresh tumour cells and cell lines. *J. Clin. Invest.* 84: 1707-1712.

CLONING AND CHARACTERIZATION OF cDNAs FOR MURINE MIP-2 AND HUMAN MIP-2 HOMOLOGS

P Tekamp-Olson,[1] C Gallegos,[1] D Bauer,[1] J McClain,[1] B Sherry,[2] M Fabre,[2] S van Deventer,[2] and A Cerami[2]. [1]Dept. of Biochemistry, Chiron Corp, Emeryville, CA 94608, USA. [2]Laboratory of Medical Biochemistry, Rockefeller University, New York NY 10021, USA

Murine MIP-2 is one of two heparin binding proteins secreted by lipopolysaccharide stimulated RAW 264.7 cells. This cytokine has been found to elicit a localized inflammatory response when injected subcutaneously into the footpads of mice, to have potent chemotactic activity *in vitro* for human polymorphonuclear leukocytes (1) and to have myelopoietic enhancing activities for CFU-GM *in vitro* (2). We have cloned and sequenced the cDNA for murine MIP-2 using degenerate oligonucleotide probe pools based on the partial N-terminal amino acid sequence of the secreted protein. We have used this cDNA to screen a library prepared from PMA and LPS-stimulated U937 polyA$^+$ RNA. We have isolated and sequenced two genes, hu-MIP-2α and hu-MIP-2β which show high homology to each other as well as to a previously described cytokine *gro*/MGSA. These three cytokines, hu-MIP-2α, hu-MIP-2β and hu-*gro*/MGSA, define a sub-family within the platelet factor 4 family of proteins.

1. Wolpe, S.D., B. Sherry, D. Juers, G. Davatelis, R.W. Yurt and A. Cerami. 1989. Identification and characterization of macrophage inflammatory protein 2. *Proc. Natl. Acad. Sci. USA.* 86: 612-616.
2. Broxmeyer, H.E., B. Sherry, L.Lu, S. Cooper, C. Carow, S.D. Wolpe, and A. Cerami. 1989. Myelopoietic enhancing effects of murine macrophage inflammatory proteins 1 and 2 on colony formation *in vitro* by murine and human bone marrow granulocyte/macrophage progenitor cells. *J. Exp. Med.* 170: 1583-1594.

ACT-2 GENOMIC STRUCTURE, CHROMOSOMAL LOCALIZATION AND IDENTIFICATION OF ACT-2 RECEPTORS

M Napolitano,[1] WS Modi,[2] VH Seuanez,[3] SJ Cevario[3], and WJ Leonard.[1] [1]Cell Biology and Metabolism, NICHD, Bethesda, 20814; [2]Biological Carcinogenesis Program and [3]Viral Carcinogenesis Lab. Frederick Cancer Research Facility, Frederick, MD 21701, USA

Act-2 is a cytokine that is expressed by PBMCs following mitogenic or anti-CD3 activation. It appears to be the human equivalent of the murine Macrophage

Inflammatory Protein (MIP)-1, a mediator of inflammation, because of its high homology (71 – 77%) with the α and β chains of this murine factor. Act-2 shares sequence similarity with a number of secreted factors, some of which play a role in inflammation. In an effort to study the mechanism of action of Act-2, we purified the recombinant Act-2 protein by HPLC and looked for the presence of receptors on a variety of cell types. We performed binding studies utilizing ^{125}I-Act-2 and identified Act-2 receptors on HeLa, K562, HL60, MT-2 and PHA/PMA-activated PBLs, but not on resting PBLs. The Scatchard plots revealed a K_d of 3-10 nM and 7 000-10 000 receptors/cell on MT – 2 and K562 cells and 45 000 receptors/cell on activated PBLs. We also isolated Act-2 genomic clones in order to study the Act-2 genomic structure and modality of gene regulation after activation. The Act-2 gene consists of three exons and contains a classical TATA box. The Act-2 promoter is functional when tested on a CAT reporter gene following PHA/PMA stimulation of Jurkat cells or when activated in trans by *tax*, an HTLV-1 gene product, analogously to a number of other cytokines and cytokine receptors. Furthermore, by using somatic cell hybrids and *in situ* chromosomal localization, we were able to assign the Act-2 gene to chromosome 17q21-q23.

Thus we have characterized regulatory sequences of the Act-2 gene and identified specific Act-2 receptors as steps to elucidate this cytokine / cytokine receptor system.

BIOLOGICAL PROPERTIES OF THE RANTES/SIS CYTOKINE FAMILY

T Schall, K Toy, and DV Goeddel. Genentech, Inc.,
460 Pt. San Bruno Blvd., S. San Francisco, CA 94080, USA

The RANTES/SIS cytokine family consists of the human leukocyte-secreted factors RANTES, pLD78, HIMAP (a variant of Act-2/hH400) and MCAF (MCP-1). This family is itself part of a larger superfamily of molecules characterized by the conserved motif of four cysteine residues which divides the superfamily into two branches: the "C-C" branch which contains the RANTES/SIS cytokines and the "C-X-C" branch which includes IL-8 and platelet factor 4. Members of the C-X-C branch have been characterized recently as neutrophil targeted inflammatory agents, but the biological activities of the RANTES/SIS cytokines are less well known. Our studies with these factors show: 1) there are distinct differences in the expression of the RANTES/SIS genes after cellular stimulation by a variety of agents. *In situ* hybridization data support the idea that these genes are differentially regulated *in vivo* as well. 2) Despite predicted structural similarities, recombinant proteins for the RANTES/SIS cytokines exhibit diverse biochemical properties. 3) In addition to MCAP, other recombinant RANTES/SIS cytokines are potent and specific chemoattractants for blood monocytes but not blood neutrophils. We have also examined the roles of these cytokines in leukocyte adhesion and monocyte activation. We suggest that RANTES/SIS cytokines may be involved in monocyte mediated inflammatory processes and in other important immune regulatory functions.

ENDOTHELIAL AND LEUKOCYTE FORMS OF INTERLEUKIN-8: CONVERSION BY THROMBIN AND INTERACTION WITH NEUTROPHILS

CA Hébert,[2] FW Luscinskas,[1] J-M Kiely,[1] EA Luis,[2] WC Darbonne,[2] GT Bennett,[2] CC Liu,[2] MS Obin,[1] MA Gimbrone Jr,[1] and JB Baker[2]
[1]Vascular Research Div., Harvard Medical School, Boston MA 02115;
[2]Genentech, Inc. SO San Francisco, CA 94080, USA.

We have recently shown that endothelial cell-derived interleukin 8 (IL-8) inhibits neutrophil adhesion to IL1β-activated human umbilical vein cell (HUVEC) monolayers (1). IL-8 secreted by T-lymphocytes or monocytes has been characterized as a promoter of neutrophil degranulation and chemotaxis (2). The IL-8 isolated from each of these cell types is a mixture of two IL-8 polypeptides, one consisting of 72 amino acids (herein called [ser-IL-8]72) and the other 77 amino acids (an N-terminal extended form herein called [ala-IL-8]77). IL-8 derived from T-lymphocytes and monocytes is predominantly [ser-IL-8]72 whereas endothelial-derived IL-8 is highly enriched ($>80\%$) in [ala-IL-8]77. The present study addresses the relationship and activities of these two forms of IL-8 using recombinant proteins expressed by both mammalian cells and *E.coli*. Thrombin was found to efficiently convert [ala-IL-8]77 to [ser-IL-8]72. In contrast, urokinase and tissue-type plasminogen activator were unable to cleave [ala-IL-8]77 and trypsin generated multiple IL-8 cleavage fragments. In competitive binding assays using ^{125}I[ala-IL-8]77, neutrophils exhibited a two-fold preference for [ser-IL-8]72 over [ala-IL-8]77. Both forms of IL-8 inhibited neutrophil adhesion to IL-1β-activated HUVEC monolayers by up to 90%. However, [ser-IL-8]72 was approximately 10-fold more potent than [ala-IL-8]77 in these assays ($EC_{50} \approx 0.3$ nM for [ser-IL-8]72 vs. ≈ 3 nM for [ala-IL-8]77). Both forms of IL-8 promoted degranulation of cytochalasin B-treated neutrophils {[ser-IL-8]72 ($EC_{50} > 10$ nM) was 2 to 3-fold more potent than [ala-IL-8]77} although, in this regard, they were less active than formylmethionyl-leucyl-phenylalanine (FMLP). These data suggest that [ala-IL-8]77 and [ser-IL-8]72 have qualitatively similar and potentially complex biological activities and that full activation of IL-8 requires cleavage to the [ser-IL-8]72 form. In the case of inflamed endothelial cells, this activation could be mediated by thrombin associated with the procoagulant environment associated with these cells.

1. Gimbrone, M.A. Jr. *et al.* 1989. *Science* 246: 1601.
2. Baggiolini, M. *et al.* 1989. *J. Clin. Invest.* 84: 1045.

INTERLEUKIN 8 ACTS ON MONOCYTES TO INCREASE THEIR ATTACHMENT TO CULTURED ENDOTHELIUM AND ENHANCE THEIR EXPRESSION OF SURFACE ADHESION MOLECULES

K A Brown,[1] F Le Roy,[1] G Noble,[1] K Bacon,[2] R Camp,[2] A Vora,[2] and D C Dumonde[1]. [1]Dept. of Immunology & [2]Institute of Dermatology, UMDS, St Thomas's Hospital Campus, London, UK.

Interleukin 8 (IL-8) is chemotactic for neutrophils and lymphocytes and is present in psoriatic skin (1,2). Monocytes comprise a significant proportion of the leucocyte infiltrates of psoriatic lesions and their extravasation into inflammatory lesions may also be influenced by the action of IL-8. To examine this possibility, we investigated if rIL-8, over the concentration range $1x10^{-7}$ to $1x10^{-11}$M, modified the behaviour of monocytes by increasing their adherence to cultured endothelium, enhancing their expression of surface adhesion molecules and/or possessing chemotactic activity.

Using a quantitative monolayer adhesion assay, rIL-8 augmented the adherence of monocytes from normal subjects to cultured umbilical cord endothelium. The optimum increase in adherence (mean 336%; $p < 0.001$) occurred when monocytes were pretreated, prior to assay, with $1x10^{-9}$ IL-8 for 4h. Such an effect was not seen when endothelial cells were pretreated with the cytokine. The increase in monocyte attachment induced by IL-8 corresponded to an enhanced expression of the adhesion molecules CD11a, CD11b and CD18 on the surface of these cells as measured by FACS analysis. Moreover, the expression of Class I antigens on the surface of monocytes was also increased following incubation of the cells with IL-8 for 4h. IL-8 was not chemotactic for monocytes though, in the same assay, we confirmed that the cytokine stimulated lymphocyte migration.

These results show that IL-8 is effective in promoting the attachment of monocytes to vascular endothelium. This hitherto unreported activity may be relevant to the recruitment of circulating monocytes into inflammatory lesions.

1. Schröder, J-M., J. Young, H. Gregory and E. Christophers. 1989. *J. Invest. Dermatol.* 92: 515.
2. Gearing, A. J. H., N. J. Fincham, C. R. Bird, M. Wadhwa, A. Meager, J. E. Cartwright, and R. D. R. Camp. 1990. *Cytokine* (in press).

NOVEL NEUTROPHIL CHEMOATTRACTANTS GENERATED IN RESPONSE TO ZYMOSAN *IN VIVO*

PD Collins, PJ Jose and TJ Williams. Dept. of Applied Pharmacology, National Heart & Lung Institute, Dovehouse Street, London SW3 6LY, UK.

Inflammatory exudate fluid recovered 2 h after intraperitoneal injection of zymosan in the rabbit contains vascular permeability-increasing activity which is neutrophil-dependent and largely due to the chemoattractant C5a (1,2). Use of neutralising anti-C5a antibodies has revealed that later samples contain neutrophil chemoattractants other than C5a.

Zymosan (500 mg) or purified rabbit C5a (60nM) for comparison, in 50 ml sterile saline containing 40μM polymyxin B to inhibit possible responses to endotoxin, was injected into the rabbit peritoneal cavity. Exudate samples, collected after 2 and 6 h, were centrifuged to remove accumulated leukocytes and any remaining zymosan particles. Inflammatory activity in the peritoneal fluids was assayed by injection into the skin of a second rabbit in the presence and absence of anti-C5a antibody and measuring the local accumulation of intravenously injected ^{125}I-albumin and ^{111}In-neutrophils. Responses were amplified using local PGE_2.

Anti-C5a antibody inhibited oedema and neutrophil accumulation in response to C5a by 100±6% and 97±6% respectively. In contrast, skin responses to zymosan-induced peritoneal exudates were not completely inhibited by the antibody. Inhibition of oedema and neutrophil accumulation respectively was 74±9% and 72±11% for 2 h exudates, 41±5% and 46±7% for 6 h exudates (mean± SEM, n=6-

8). All activity in C5a-induced exudate was totally inhibited by the anti-C5a antibody. Cation exchange and reverse phase HPLC of 6 h exudates induced by zymosan and depleted of C5a using affinity columns revealed two peaks of activity. One of these peaks contained a peptide showing homology to human IL-8.

These results demonstrate that, in response to a specific inflammatory stimulus *in vivo*, there is the sequential generation of several neutrophil attractants that may act in concert to regulate the inflammatory response.

1. Jose, P.J., M.J. Forrest, and T.J. Williams. 1983. *J. Exp. Med.* 158: 2177–2182.
2. Forrest, M.J., P.J. Jose and T.J. Williams. 1986. *Br. J. Pharmacol.* 89: 719–730.

DIFFERENT PRO-INFLAMMATORY PROFILES OF MACROPHAGE DERIVED INTERLEUKIN-1, -8 AND TUMOUR NECROSIS FACTOR

M Rampart,[1] J Van Damme,[2] W Fiers,[3] and AG Herman.[1]
[1]Division of Pharmacology, University of Antwerp; [2]Rega Institute Leuven; [3]State University of Ghent, Belgium.

In vitro, IL-1, IL-8 and TNF have biological properties (e.g. increased adhesiveness of endothelial cells, activation of neutrophils) which imply them as putative mediators of inflammation. In order to explore their pro-inflammatory properties *in vivo*, IL-1, IL-8 and TNF were injected intradermally (id) in the clipped dorsal skin of anaesthetized rabbits. Neutrophil infiltration and plasma leakage were measured as the local accumulation of intravenously (iv) injected ^{111}In-neutrophils and ^{125}I-albumin. When injected alone, these cytokines induced only very small responses. When local blood flow was enhanced by co-injection of prostaglandin E_2 (PGE_2, 300 pmol) or calcitonin gene-related peptide (CGRP, a peptide vasodilator, 10 pmol), human and mouse recombinant TNF (0.3-30 fmol), human authentic IL-8 (0.02 – 20 pmol) and human recombinant IL-1 (α and β, 0.03 – 3 fmol) were all very potent at inducing neutrophil accumulation. Cell filtration in response to TNF and IL-8 was fast in onset and short of duration with a half-life of 6 – 8 min for TNF and 60 – 70 min for IL-8. IL-1-induced neutrophil accumulation was slow in onset, required 4 – 6h to reach its maximum and was protein biosynthesis dependent. Furthermore, TNF- and IL-8-induced neutrophil emigration was associated with a parallel time course of plasma leakage, whereas IL-1 did not induce any significant oedema formation (even in the presence of PGE_2 or CGRP). These differences in time course and biological profile indicate that IL-1, IL-8 and TNF exert their pro-inflammatory effects *in vivo* via different mechanisms. Our data also suggest that tissue macrophages can initiate an inflammatory response by signalling the presence of an injurious stimulus to the blood vessels by means of cytokines such as IL-8 and TNF. IL-1 may be important for the sustained attraction of neutrophils in the later stages of inflammation. Finally, β-thromboglobulin (β-TG, 2 – 20 pmol), a platelet release product with 44% sequence homology with IL-8, also induced neutrophil emigration and plasma leakage (in the presence of PGE_2 or CGRP). Thus, via release of β-TG, platelets may contribute to neutrophil activation during inflammatory reactions.

INTERLEUKIN-8 ENHANCES THE INTRACELLULAR KILLING OF *MYCOBACTERIUM FORTUITUM* BY HUMAN GRANULOCYTES

O Pos, MF Geertsma, A Stevenhagen, PN Nibbering, and R van Furth. Dept. of Infectious Diseases, University Hospital, Leiden 2300 RC Leiden, The Netherlands.

The cellular defense mechanisms against mycobacteria have not been entirely elucidated. Although the activated macrophage is regarded as the most important effector cell in the elimination of these bacteria, most studies have failed to show the intracellular killing of mycobacteria by these cells *in vitro*. Cytotoxic T-lymphocytes can lyse macrophages infected with mycobacteria, which results in the appearance of extracellular bacteria. This has led us to investigate the ability of human granulocytes to phagocytose and intracellularly kill mycobacteria *in vitro* and whether the microbicidal functions of these cells can become activated by the cytokines Interleukin-8 (rIL-8) and Interferon-gamma (rIFN-gamma). *M.fortuitum* was chosen as an example of mycobacteria because of its rapid growth and its relatively little tendency to clump.

Human granulocytes were found to phagocytose opsonized *M.fortuitum* very rapidly, but showed a rather poor intracellular killing of the ingested bacteria. Pre-incubation of the granulocytes with rIFN-gamma for 18h resulted in an increased intracellular killing of opsonized *M. fortuitum*, whereas phagocytosis was not affected. The enhanced killing was not accompanied by an increase in the extracellular release of H_2O_2 upon stimulation with PMA.

Since IL-8 was reported to enhance the growth inhibitory capacity of human granulocytes for *Candida albicans*, we were interested to find out whether this more granulocyte-specific cytokine could also influence the phagocytosis and intracellular killing of opsonized *M.fortuitum* by human granulocytes. The presence of more than 9 nM rIL-8 during the killing assay resulted in an increased intracellular killing of *M.fortuitum*. rIL-8 did not affect the phagocytosis of opsonized *M.fortuitum* by human granulocytes. The PMA-induced release of extracellular H_2O_2 was found to be stimulated by rIL-8.

In conclusion, we have shown for the first time that rIL-8 stimulates the microbicidal capacity of human granulocytes.

INHIBITORS OF PROTEIN KINASE C ACTIVATION INHIBIT INTERLEUKIN-8-INDUCED *IN VITRO* HUMAN LYMPHOCYTE MIGRATION

KB Bacon and RDR Camp. Institute of Dermatology, St. Thomas's Hospital, Lambeth Palace Road, London SE1 7EH, UK.

The mechanism of the *in vitro* lymphocyte chemoattractant action of human recombinant (hr) interleukin (IL) 8 has been further characterised by assessing the role of protein kinase C (pkC) inhibition and activation.

Migration of mixed peripheral blood lymphocytes (PBL) was assessed using a 48-well migration assay, results being expressed as a migration index (MI; area of the lower surface of the filter occupied by cells in response to chemoattractant [mm^2]/area of filter occupied by unstimulated cells [mm^2]. The effect of hrIL-8 on pkC activation was assessed by preincubating the PBL with the pkC inhibitors H7 ($10^{-12} - 10^{-7}$M) and sphingosine ($10^{-7} - 10^{-4}$M) for 10 minutes at 37°C and then assessing migration in response to hrIL-8 (10^{-11}M). In three separate experiments, the effect of preincubating PBL for 30 minutes in the presence of the direct pkC activators 1-oleoyl-2-acetyl-sn-glycerol (OAG) and 1,2-dioctanoyl-sn-glycerol (DOG; $10^{-8}-10^{-4}$M) and the inhibitors H7 (10^{-8}M) and sphingosine (10^{-5}M) on hrIL-8-induced PBL migration, was assessed.

The mean MI±SEM in the presence of hrIL-8 (10^{-11}M) alone was 1.97±0.14, n=8. H7 completely inhibited the hrIL-8-induced PBL migration over the dose range $10^{-12}-10^{-7}$M (n=5). The unstimulated migration (response to medium alone) was unaffected over this dose range. Sphingosine inhibited hrIL-8-induced PBL migration in a dose-dependent manner between 10^{-7} and 10^{-4}M (the IC_{50} value being approximately 8 x 10^{-7}M; n=4), without affecting unstimulated migration. In the presence of the pkC activators OAG and DOG, the inhibitory activity of a 30 minute preincubation with H7 on hIL-8-induced PBL migration was partially reversed (MI being 1.52 and 1.47 at 10^{-5}M for OAG and DOG respectively, n=3). Similar results were obtained when PBL were pre-incubated for 30 minutes in the presence of sphingosine and either OAG or DOG (MI being 1.69 and 1.44 at 10^{-5}M, for OAG and DOG respectively, n=3). OAG and DOG ($10^{-8}-10^{-4}$M) alone had no effect on unstimulated lymphocyte migration.

IL-8-induced *in vitro* PBL migration therefore appears to be dependent on pkC activation, as shown by the inhibition of the migratory response by the inhibitors H7 and sphingosine, the lack of effect of inhibitors on unstimulated migration and the partial reversal of these effects by the activators OAG and DOG. The lack of complete reversal of the inhibitory effects can in part be explained by the non-specific inhibitory activity of H7 and sphingosine, however pkC activation may be an integral part of the signal transduction process during IL-8-stimulated human PBL migration.

TUMOUR NECROSIS FACTOR (TNFα), THE INFLAMMATORY COMPONENT OF HIV-RELATED LUNG DISEASE?

AB Millar, A Meager, SJG Semple, and GAW Rook. Depts. of Medicine and Medical Microbiology, University College & Middlesex School of Medicine, London and NIBSC, S Mimms, UK.

Tumour necrosis factor (TNF) induces many physiological and biochemical changes in man which can be broadly considered as part of host defence. A role for TNF in the pathogenesis of infection with the human immunodeficiency virus (HIV) has been postulated (1). The lung is often affected in HIV infection and is the site of the commonest opportunistic infection, pneumocystis carinii pneumonia (PCP). Some of these patients develop respiratory failure and we postulated that TNF release might be involved, as has been found in other causes of respiratory failure in non-HIV positive patients (2).

We investigated 20 HIV positive patients with respiratory symptoms who underwent fibreoptic bronchoscopy (FOB) and bronchoalveolar lavage (BAL). The BAL supernatant from all patients was stored. Ten patients had PCP, five had bacterial pneumonia (BP) and five had normal bronchoscopic findings (NB). Peripheral blood monocytes were obtained by venepuncture before FOB. Monocytes and macrophages were harvested by adherence to plastic. The cells were cultured with RPMI (+10% serum) and with sequential dilutions of LPS. The same procedure was followed in eight seronegative controls (C). TNF was assayed in the BAL fluid and the cell culture supernatants using a double sandwich ELISA technique. The spontaneous production of TNF from the alveolar macrophages for each patient group was (mean±SD iu/ml) 205±57 (PCP), 173±32 (BP), 112±47 (NB), 11±9 (C) and for peripheral blood monocytes was 124±25 (PCP), 111±46 (BP), 89±34 (NB) and 13±8 (C). The maximally stimulated levels of TNF from alveolar macrophages were (mean±SD iu/ml) 259±73 (PCP), 235±35 (BP), 215±51 (NB) and 61±12 (C) and those from peripheral blood monocytes were 203±48 (PCP), 170±58 (BP), 65±29 (NB), 61±12 (C).

We conclude that alveolar macrophages and peripheral blood monocytes from HIV infected individuals secrete TNF more readily that those from normal controls. In addition, there is an increase in TNF production with more severe forms of lung disease. Whether this is due to intrinsic pulmonary pathology or related to the degree of HIV infection remains speculative.

1. Vyakarnam *et al. AIDS*. 1990. 4: 21 – 27.
2. Millar *et al. Lancet*. 1989. 2: 713 – 715.

MACROPHAGE HETEROGENEITY AND TUMOUR NECROSIS FACTOR-α RELEASE

M Stein and S Gordon. Sir William Dunn School of Pathology
Oxford University, Oxford, OX1 3RE, UK

TNFα is an important mediator in host defense function, yet little is known about the rules governing its release from macrophages. This study examines (a) the role of recruitment in priming macrophages for TNFα release and (b) the role of important phagocytic receptors, Fc receptors (FcR) and complement receptors, in mediating release of TNFα from differentially recruited populations of mouse peritoneal macrophages.

BCG-, biogel bead- and thioglycollate-elicited (TPM) populations of macrophages were incubated with one of: LPS, zymosan, latex particles or sheep erythrocytes (SE) opsonized with either IgG or complement. After a 6 hour incubation in the continuous presence of the above stimuli, the medium was removed and an aliquot assayed by L929 fibroblast bioassay. Results: 1) LPS challenge of macrophages harvested 4 to 6 days following i.p. BCG infection release small amounts (24 U/ml) of TNFα whereas macrophages harvested 10 to 14 days following infection consistently release more than 256 U/ml; 2) LPS triggered the release of large amounts of TNFα from TPM but not from biogel-elicited macrophages (BgPM); yet both macrophage populations release similar amounts of TNFα (about 500 U/ml) following challenge with zymosan and IgG – opsonized SE. The release following IgG opsonized SE was blocked in the presence of 2.4G2 (a rat anti-mouse Fc receptor for IgG1 and 2b monoclonal antibody). IgM-complement opsonized SE does not trigger release of any TNFα from BgPM and only barely detectable amounts from TPM. Neither latex nor gluteraldehyde fixed SE trigger release of TNFα in any of the above macrophage populations. The data were confirmed by western blot analysis.

The results suggest that: (a) recruitment *per se* is not sufficient to prime macrophages for LPS-induced TNFα release; (b) there are at least two pathways mediating TNFα release since BgPM are not sensitive to LPS, yet release significant amounts of TNFα following phagocytic triggering; (c) the FcR for IgG1/2b, mediates TNFα release, in contrast to the barely detectable amounts measured following complement receptor ligation.

CONDITIONS FOR THE DEMONSTRATION OF TUMOUR NECROSIS FACTOR mRNA IN REACTIVE AND NEOPLASTIC CELL POPULATIONS

K Morrison and DB Jones. University Pathology,
General Hospital, Southampton SO9 4XY, UK

We have undertaken an *in situ* hybridisation study of Hodgkin's lymphoma and reactive macrophage populations in sarcoid and tuberculosis. The system we have employed consists of synthetic oligonucleotide probes labelled with biotin 3 and 5. Specific hybridisation to tissue sections is demonstrated using an alkaline phosphatase technique. The use of multiple oligonucleotide cocktails, together with proteinase K pretreatment, enables the demonstration of mRNA sequences in routine formalin fixed, paraffin embedded tissue with good resolution and morphology. Studies of giant cell and epithelioid macrophages in tuberculosis and sarcoid, using probes directed to the poly-A tails of mRNA demonstrate that these cells are highly active in terms of protein synthesis. Specific probes directed to TNFα have shown hybridisation to TNF message sequences. Hybridisation is present in giant cells and epithelioid macrophages and the signal is digestible with RNase. In Hodgkin's lymphoma, where the pathological features suggest heavy recruitment of inflammatory cells and extensive fibrosis, an RNase digestible TNFα signal is present within the tumour cell population. These studies, which have been confirmed by parallel staining with specific antibody to TNFα, suggest a significant role for this cytokine in the pathogenesis of Hodgkin's disease. The procedures described, including the use of parallel demonstration of cytokine message and protein using hybridisation and antibody staining techniques, represent a powerful tool for the demonstration of pathogenetic mechanisms in complex histological conditions in routine biopsy material.

THE STRUCTURE OF TUMOUR NECROSIS FACTOR

EY Jones,[1] DI Stuart,[1] and NPC Walker [2]
[1]Molecular Biophysics, University of Oxford, Oxford OX1 3QU, UK.
[2]BASF AG, Ludwigshafen, FRG

We have determined the three dimensional structure of TNF at 2.9Å using the technique of X-ray crystallography (1). The structure has been refined to a current R factor of 20% with excellent stereochemistry hence the individual atomic coordinates are accurate to within 0.3Å. Such a structure provides a wealth of information, not only on specific atom positions but, in addition, their interactions (H bonds etc), solvent accessibility and variations in flexibility/mobility within the molecule.

TNF is a compact trimer composed of three identical subunits of 157 amino acids. The main-chain topology for a single subunit is essentially a β-sandwich structure formed by two anti-parallel β-pleated sheets. The subunits associate tightly about the threefold interacting through a simple edge to face packing of the β-sandwich to form the solid, conical-shaped trimer. The mainchain fold of the monomer corresponds to the 'jelly roll' motif observed in viral coat proteins such as VP1, VP2 and VP3 of rhinovirus or the hemagglutinin molecule of influenza. TNF is the first non-viral protein to contain this motif. It is possible that this structural homology will lead to surprising insights into the function of both TNF and the viral proteins.

The relationship between structure and function for TNF may now be explored using a wealth of published data on antibody binding, neutralising antibodies, site directed mutagenesis and the properties of the homologous molecule, lymphotoxin. In particular, very recent data from extensive site directed mutagenesis experiments indicate the probable location and nature of the receptor binding site.

1. Jones, E.Y., D.I. Stuart, and N.P.C. Walker. 1989. Structure of Tumour Necrosis Factor. *Nature.* 338: 225–228.

IL–8 IS A POTENT STIMULATOR OF NEUTROPHIL AGGREGATION AND CHEMOTAXIS IN CANINE AND HUMAN NEUTROPHILS

MK Thomsen, CG Larsen, K Thestrup–Pedersen. LEO Pharm. Corp., Ballerup; Dept. of Dermatology, University of Aarhus, Denmark.

We have studied and compared the *in vitro* effect of IL–8 and the lipid mediators LTB_4 and PAF on human or canine neutrophils aggregation and chemotaxis. We found, that all three mediators induced the *in vitro* aggregation of neutrophils in both species. IL–8 induced more rapid aggregation than PAF and more persistent aggregation than LTB_4. The potency of IL–8 with respect to neutrophil aggregation as well as chemotaxis was 100 times higher than that of the lipid mediators. A surprising finding was that the maximal chemotactic response (E_{max}) to IL–8 was significantly greater than E_{max} for LTB_4 and PAF in canine, but not human, cells. In the context of IL–8 production, we have recently described a novel inhibitor of leukotriene synthesis, ETH615, which, in an equipotent manner, inhibits production of IL–8 and LTB_4 in human monocytes and neutrophils.

In conclusion, IL–8 is an extremely potent stimulator of neutrophil aggregation and chemotaxis, *in vitro*, being in the order of 100 times more potent than the classical chemoattractants, LTB_4 and PAF. Furthermore, the dog appears to represent a highly responsive model for studies of IL–8.

ENDOGENOUS AND SYNTHETIC INHIBITORS OF THE PRODUCTION OF CHEMOTACTIC CYTOKINES

CG Larsen, M Kristensen, K Paludan, B Deleuren, MK Thomsen, K Kragballe, K Matsushima, and K Thestrup – Pedersen
Dept. of Dermatology & Institute of Molecular Biology, University Aarhus; LEO Pharmaceutical Corp., Ballerup, Denmark. Laboratory of Molecular Immunoregulation, NIH, Frederick MD, USA.

IL–8 is believed to be involved in the pathogenesis of psoriasis as well as rheumatoid arthritis. IL–8 is markedly inducible by IL–1 and TNFα and is responsible for some of the pro–inflammatory activities of these cytokines. The putative role of IL–8 as a key mediator in inflammation underscores the need to identify regulatory mechanisms of IL–8 production. Corticosteroids have been shown to be potent inhibitors of IL–8 production. Furthermore, we have demonstrated an inhibitory effect of another endogenous steroid molecule,

1,25-$(OH)_2$-vitamin D_3 (D_3). D_3 inhibits the *in vitro* expression of IL-8 mRNA (as well as IL-1 mRNA) in LPS-stimulated human mononuclear cells at pharmacological relevant concentrations ($10^{-8}-10^{-11}$M). These results suggest that D_3, a potent anti-psoriatic drug (Kragballe 1989), may act as an important endogenous regulator of IL-1 and IL-8. Also, we identified ETH615, an inhibitor of leukotriene-biosynthesis, as a potent suppressor of IL-8 (and IL-1) mRNA expression in LPS-stimulated human mononuclear cells. ETH615 inhibits zymosan-induced cutaneous inflammation in dogs (Thomsen, 1990) and may be a nonsteroid alternative in the treatment of inflammatory skin disorders.

INDUCTION OF HAPTOTACTIC MIGRATION OF MELANOMA CELLS BY NEUTROPHIL ACTIVATING PROTEIN/IL-8

JM Wang,[1] G Taraboletti,[1] K Matsushima,[2] J Van Damme,[3] and A Mantovani.[1] [1]Istituto di Richerche Farmacologiche Mario Negri, Milan, Italy. [2]Laboratory of Molecular Immunoregulation, NCI, Frederick, USA. [3]Rega Institute for Medical Research, University of Leuven, Belgium.

IL-8 belongs to an emerging group of cytokines involved in inflammation. While it is clear that several of these cytokines are involved in the regulation of leukocyte recruitment and activation, the spectrum of action of individual members of this group of peptide mediators remains to a large extent to be defined.

In this study, we report that IL-8 induces haptotactic migration of human melanoma cells. When seeded in the lower compartment of the modified Boyden chamber, natural and recombinant IL-8 induced migration across polycarbonate filters of human A2058 melanoma cells. Checkerboard experiments revealed a gradient-dependent response of A2058 melanoma cells to IL-8. Anti-IL-8 antibodies blocked IL-8-induced melanoma cell migration, indicating that the IL-8 molecule is indeed responsible for the induction of melanoma cell chemotaxis. Polycarbonate filters exposed to IL-8 for 20h and washed, caused migration of melanoma cells, thus demonstrating a haptotactic response of melanoma cells to IL-8. The homologous polypeptide platelet factor 4 was inactive in inducing melanoma cell chemotaxis or haptotaxis.

The observation that IL-8 affects melanoma cells emphasizes the need for a comprehensive analysis of the spectrum of action of platelet factor-4 related peptides. The effect of the inflammatory cytokine IL-8 on melanoma cells may be relevant to augmented secondary localization of tumours at sites of inflammation.

CHEMOTACTIC CYTOKINE PRODUCTION BY VASCULAR CELLS

A Sica,[1] JM Wang,[1] J Van Damme,[3] K Matsushima,[2] K Zachariae,[2] CG Larsen,[2] and A Mantovani.[1] [1]Istituto Mario Negri,Milan, Italy. [2]Laboratory of Molecular Immunoregulation, NCI, Frederick, USA. [3]Rega Institute for Med. Research, University of Leuven, Belgium.

Leukocytes and vascular cells interact closely in inflammation and immunity and cytokines are important mediators of this interaction. The present study was designed to define the capacity of human vascular cells (endothelial cells, EC, and smooth muscle cells, SMC) to produce chemotactic cytokines. Upon exposure to LPS, IL-1 or TNF (but not IL-6), EC released polypeptide chemoattractants active on monocytes and neutrophils. Concomitantly, activated EC expressed mRNA transcripts for monocyte chemotactic protein (MCP) and IL-8/NAP-1. The latter finding confirms a previous observation by Strieter *et al.*(1) Nuclear run off analysis revealed that IL-1 activated transcription of the MCP and IL-8/NAP-1 gene. Unlike EC supernatants, conditioned media from vascular SMC exposed to inflammatory signals induced selectively monocyte migration, with no effect on PMN. Concomitantly SMC expressed MCP mRNA transcripts. The production of chemotactic cytokines may represent one of the mechanisms whereby vascular cells, exposed to inflammatory signals, participate in the regulation of leukocyte extravasation.

1. Strieter *et al.* 1989. *Science.* 243: 1479.

IN VIVO INFLAMMATORY ACTIVITY OF NEUTROPHIL-ACTIVATING PEPTIDE-1

I Colditz[1] and M Baggiolini.[2] [1]CSIRO Division of Animal Health Armidale 2350, Australia. [2]Theodor Kocher Institute, University of Berne, CH-3000 Berne, Switzerland.

The inflammatory activity of recombinant human neutrophil activating peptide-1 (NAP-1/interleukin-8) was examined in skin of conscious rabbits. ^{51}Cr-labelled neutrophils and ^{125}I-labelled human serum albumin were used to quantify neutrophil accumulation and plasma leakage induced by rhNAP-1, rhIL-1, FMLP and endotoxin.

NAP-1 was of comparable potency to FMLP with ≥1 picomole/site causing neutrophil accumulation. Co-injection of 300 picomoles PGE_2 caused approximately threefold increase in neutrophil accumulation and tenfold increase in plasma leakage for a range of NAP-1 doses. Significant neutrophil accumulation continued for at least 8 hours, in contrast to the more transient accumulation of neutrophils induced by endotoxin, IL-1 and classical chemotactic agonists. Plasma leakage was found to be dependent on the presence of circulating neutrophils and declined more rapidly than did neutrophil accumulation as lesions aged. Inflammatory activity of NAP-1 was not affected by co-injection of the protein synthesis inhibitor actinomycin D (AD, 50 nanomoles/site) which, however, profoundly suppressed inflammatory actions of endotoxin. We postulated that an endothelial response may be initiated to limit plasma leakage during ongoing neutrophil emigration at sites stimulated with NAP-1. AD did not abrogate the decline in plasma leakage, thus *de novo* protein synthesis is unlikely to contribute to the mechanism that restricts plasma leakage in older inflammatory lesions. When extant lesions were restimu-ated with NAP-1, neutrophil accumulation and plasma leakage were diminished.

NAP-1 shares with classical chemotactic agonists similar potency, independence from *de novo* protein synthesis to exert an inflammatory effect and dependence of plasma leakage on neutrophils; the long duration of action and moderate degree of homologous desensitization induced by NAP-1 indicate that it may be a major mediator of inflammation in diverse pathologies.

IL-8 PRODUCES INFLAMMATORY HYPERALGESIA VIA A SYMPATHETIC COMPONENT

FQ Cunha, BB Lorenzetti, and SH Ferreira. Dept. of Pharmacology, Faculty of Medicine of Ribeirao Preto, University of Sao Paulo, 14049 Ribeirao Preto, SP, Brazil.

Inflammatory hyperalgesia has two basic components, one mediated by cyclo-oxygenase metabolites, the other by the release of sympathetic amines. The first component is blocked by aspirin-like drugs and the second by betablockers or DA-1 antagonists. Both components are induced by carrageenin in the rat paw and measured by a modification of the Randall-Selitto test. The cyclo-oxygenase component seems to be secondary to the release of IL-1, since it is abolished by its specific antagonist, LYS-(D)-PRO-TYR. It was not know if the sympathetic component also involved mediation by a cytokine. In the present communication, we studied the ability of human purified IL-8 to cause hyperalgesia in the same nociceptive model.

It was found that IL-8 caused a dose-dependent hyperalgesia which, in contrast to that of IL-1 was not blocked by indomethacin (100 μg/paw) and was abolished by treatment of the animals with a betablocker (propranolol, 50 μg/kg or atenolol, 25 μg/paw) or a sympatholytic, guanethidine (30 mg/kg, on three consecutive days).

These results show that IL-8 caused hyperalgesia by the release of sympathetic amines and raise the possibility that IL-8 mediates sympathetic hyperalgesia. The development of antagonists for IL-8 might constitute a specific analgesic for sympathetic pain.

INTERLEUKIN-4 INHIBITS THE EXPRESSION OF INTERLEUKIN-8 FROM STIMULATED HUMAN MONOCYTES

TJ Standiford,[1] RM Strieter,[1] SW Chensue,[2] J Westwick,[3] K Kasahara[2] and SL Kunkel.[2] [1]Depts. of Internal Medicine and [2]Pathology, University of Michigan Med.School, Ann Arbor, USA. [3]Dept. of Pharmacology, Royal College of Surgeons, London, UK.

Peripheral blood monocytes are important mediators of inflammation via the generation of various bioactive substances including the recently isolated and cloned chemotactic peptide, interleukin 8 (IL-8). Through cytokine networking, monocyte-derived cytokines are capable to inducing IL-8 expression from non-immune cells. Interleukin 4 (IL-4), a B and T lymphocyte stimulatory factor, has recently been shown to inhibit monocyte/macrophage function, including the ability to suppress monocyte-generated cytokines.

We now describe the *in vitro* inhibition of IL-8 gene expression and synthesis from LPS-, TNF- and IL-1-stimulated peripheral blood monocytes by IL-4. IL-4 suppressed IL-8 gene expression, as assessed by Northern blot analysis, from stimulated monocytes in a dose-dependent fashion, with partial suppression observed at IL-4 concentrations as low as 10 pg/ml. The IL-4-induced suppressive effects were observed even when IL-4 was administered 2 hr post LPS-stimulation. Additionally, IL-4 inhibited the generation of antigenic IL-8 from LPS-, TNF- and IL-1-stimulated monocytes by 89, 88 and 96%, as assessed by ELISA. The IL-4-induced inhibition of IL-8 mRNA expression was dependent upon protein synthesis as the suppressive effects of IL-4 were significantly negated by the addition of cycloheximide.

Our findings suggest that IL-4 may be an important endogenous regulator of inflammatory cell recruitment, and add further support to the potential role of IL-4 as a down-regulator of monocyte immune function.

IN VIVO NEUTROPHIL MIGRATION INDUCED BY TUMOR NECROSIS FACTOR, INTERFERON-GAMMA AND INTERLEUKINS 1 OR 8

FQ Cunha, RA Ribeiro, LH Faccioli, GEP Souza, CA Flores and S H Ferreira. Dept. of Pharmacology, Faculty of Medicine of Ribeirao Preto, Univ. Sao Paulo, 14049 Ribeirao Preto, Brazil.

The α and β forms of recombinant interleukin-1 (IL-1α; 4.5 – 135 fmoles and IL-β; 1.5 – 13.5 fmoles), purified human interleukin-8 (IL-8; 3 – 6 fmoles) and recombinant tumor necrosis factor (TNFα; 1.5 – 750 fmoles and TNFβ; 0.6 – 300 fmoles) and interferon-gamma (IFN-gamma; 1.4 – 4.2 fmoles) induced a dose-dependent neutrophil migration into rat peritoneal cavities.

Pretreatment of the animals with dexamethasone (0.5 mg/kg) or depletion of the peritoneal resident cell population, abolished the neutrophil migration induced by those cytokines. IL-1α or IL-1β, IFN-gamma and TNFα induced neutrophil migration was increased in thioglycollate-treated rats. In contrast, neutrophil migration induced by IL-8 was not enhanced with the increase in peritoneal macrophage population.

In vitro stimulation of macrophage monolayers with IL-1β, TNFα and IFN-gamma released a factor into the supernatant which, unlike these cytokines, induced neutrophil migration in dexamethasone pretreated animals. In the rat air-pouch model, IL-8, in contrast with other cytokines studied, was unable to induce neutrophil migration. However, IL-8 was able to induce neutrophil migration after transference of homologous peritoneal cells (but not T-lymphocytes) to air-pouch. Thus, though macrophages are not associated with neutrophil migration induced by IL-8, this response seems to be dependent on a non defined peritoneal cell.

In conclusion, our results indicate that neutrophil migration induced by all cytokines studied is not due to a direct chemotactic effect on neutrophils but occurs via the release of a chemotactic factor(s) from resident cells. It is assumed that dexamethasone blocks its release.

ADHERENCE TO PLASTIC INDUCES IL-8 EXPRESSION FROM MONONUCLEAR PHAGOCYTES

K Kasahara,[1] RM Strieter,[2] SW Chensue,[1] TJ Standiford,[2] and SL Kunkel.[2] [1]Depts. of Pathology and [2]Internal Medicine, University of Michigan Medical School, Ann Arbor, Michigan USA

The hallmark of acute inflammation is the presence of adherent inflammatory cells in the microvasculature. This adherent stage of extravasation of inflammatory cells is an essential component to the initiation of cellular migration. Previous investigations have shown early gene expression with adherence to plastic for IL-1, TNF and CSF-1.

We demonstrate that human mononuclear cells adherent to plastic induce early gene expression for IL-8 as compared to cells cultured in teflon chambers. The expression of mononuclear cell-derived IL-8 by plastic adherence was shown to be both protein- and time-dependent, as assessed by Northern blot, immunohistochemical and chemotactic analyses. Adherence-induced IL-8 mRNA was inhibited with co-culture with either cycloheximide (5µg/ml) or Actinomycin D (5µg/ml). Kinetic analysis of plastic-induced, steady-state levels of IL-8 mRNA was maximal at 8 h as compared to only 20% induction by teflon.

Immunohistochemical staining of mononuclear cells on plastic, using specific rabbit anti-human IL-8 antibody demonstrated significant immunolocalization of cell-associated IL-8 antigen at 2 h, peaking at 8 h, with persistence over the next 16 h. In contrast, mononuclear cells cultured in teflon only expressed a fifth of the activity achieved with plastic adherence. Chemotactic bioactivity paralleled immunohistochemical presence of IL-8.

These data support the essential role of adherence for mononuclear cell-derived IL-8 production, which may be important for the subsequent elicitation of neutrophils into sites of inflammation.

MONOCYTE CHEMOATTRACTANTS IN STRATUM CORNEUM FROM PSORIATIC LESIONS

DG Quinn and RDR Camp. Institute of Dermatology, St Thomas's Hospital, Lambeth Palace Road, London SE1 7EH, UK.

Infiltrates of monocyte/macrophages are a characteristic histological feature of the lesions of the common inflammatory and proliferative skin disease, psoriasis. To investigate factors that may be involved in the recruitment of these cells, aqueous extracts of stratum corneum from untreated lesions ($n=10$) and from normal heel as control ($n=5$) were tested for human peripheral blood monocyte attractant activity using a 48-well microchemotaxis assay. Mixed peripheral blood mononuclear cells from normal volunteers were prepared by density gradient centrifugation on Ficoll and added to the upper wells of the chemotaxis chamber ($2x10^6$ cells ml^{-1}) which were separated from the test substance in the lower wells by a 5µm pore-size polycarbonate membrane. After a 30 min incubation, the migration of monocytes through the membrane was quantitated by image analysis of cells adherent to the membrane undersurface.

Under these conditions, no lymphocyte migration was observed. Dilution-related activity was detected in supernatants from both psoriatic and normal heel extracts, maximal migration indices (MMI) being significantly greater in the psoriatic than in control samples (5.54±1.05 versus 2.5±0.37, mean±SE; $p<0.05$, t-test). Ultrafiltration of the aqueous psoriatic stratum corneum supernatants through YM10 membranes and bioassay of filtrates and retentates revealed greater activity in the >10kD fractions than in the <10kD fractions (MMI 4.28±0.86 versus 1.63±0.31; $p<0.01$, t-test). Reversed-phase HPLC of >10kD samples and assay of 1 min fractions consistently revealed two peaks of monocyte attractant activity, eluting at 60% and 100% acetonitrile respectively ($n=5$). Further purification of the earlier eluting reversed-phase HPLC peak by TSK size exclusion HPLC, and bioassay of fractions, revealed a peak of activity eluting with a relative molecular mass of less than 13.7kD, at least 3 min later than standard $C5a_{des\ arg}$.

Analysis of the recombinant cytokines, interleukin (IL)-1, IL-3, IL-4, IL-6 and IL-8, showed that they elicited minimal or no responses in the monocyte migration assay. Previously uncharacterized monocyte chemoattractant activity in psoriatic lesional samples has been shown to be due to at least two components, which are different from $C5a_{des\ arg}$ and interleukins 1, 3, 4, 6 and 8. The precise identity, source and specificity to psoriasis of the two, partially purified, biologically active factors remain to be determined but they could play a central role in the induction of lesional monocyte infiltration in this disease.

INHIBITION OF IL-8 BY PROTEIN KINASE C (pkC) INHIBITORS IN HUMAN KERATINOCYTES

A Haslberger, C Foster, M Ceska, N Ryder, E Kugler and I Lindley
Sandoz Forschungsinstitut, Brunnerstr.59, 1235 Vienna, Austria.

Infiltration of leucocyte subpopulations into the epidermis is seen in various skin diseases. Recently, stimulation of the synthesis of the chemotactic neutrophil activating protein (NAP1/IL-8) in keratinocytes was reported. We therefore investigated the regulation of IL-8 synthesis in the basal human keratinocyte-like HACAT cell line using northern blots and a highly sensitive double ligand ELISA. A dose-dependent expression of IL-8 protein (0.04 - 2 ng/ml) was found after stimulating cells with IL-1α (250, 25 U/ml) and TNFα (1000, 100, 10 U/ml). This was accompanied by a marked accumulation of IL-8 mRNA. LPS, PMA and bradykinin did not induce IL-8 synthesis although they induced c-fos expression. In non toxic doses, the pkC inhibitors staurosporine (3 nM) and calphostin C (0.1 - 0.6μM) clearly inhibited the TNFα-induced IL-8 synthesis, but not keratinocyte proliferation. A markedly decreased IL-8 mRNA accumulation due to calphostin C showed that this inhibition is at least partially due to control at the transcriptional level. Down-regulation of pkC by pretreatment of cells with phorbol esters decreased the TNF-induced IL-8 release. IL-8 synthesis was furthermore inhibited by dexamethasone whereas cyclosporin A treatment and PGE_2 had no inhibitory effects.

These results suggest that steroids and pkC inhibitors may interfere with the synthesis of molecules controlling the trafficking of leukocytes in inflammatory processes.

MONOCYTE CHEMOTAXIS AND ACTIVATING FACTOR PRODUCTION BY CYTOKINE-ACTIVATED HUMAN EPIDERMAL KERATINOCYTES

JNWN Barker, ML Jones, RS Mitra, C Swenson, K Johnson, JC Fantone, VM Dixit and BJ Nickoloff. Depts. of Dermatology and Pathology, University of Michigan Med. School, Ann Arbor, USA.

Monocytes accumulate in the epidermis and along the dermo-epidermal junction in inflammatory skin diseases such as psoriasis and graft-versus-host disease. They are thought to play a central initiating role in the pathophysiology of these diseases, through their production of many pro-inflammatory molecules, including the cytokines TNFα and IL-1 which profoundly affect many different cutaneous cell types, including human epidermal keratinocytes (HEKs). Since HEKs, in response to IFN-gamma, express ICAM-1, and adhesive ligand for LFA-1-expressing monocytes, we sought to determine if HEKs also elaborated the specific monocyte chemotaxin and activating factor (MCAF). HEKs were cultured in a serum-free, low-calcium medium to which was added various stimulating factors. At 18 hours, RNA was extracted from the cell layer and probed with a 900 base cDNA fragment recognizing MCAF mRNA (1kb). Constitutive expression of MCAF mRNA by HEKs was noted; this could not be abolished by 24 hr incubation in basal medium or by the addition of TGFβ. MCAF mRNA was up-regulated by IFN-gamma (100 U/ml) and superinduced by IFN-gamma and TNFα (250 U/ml) in combination. TNFα alone, IL-β and phorbol esters had no effect on HEK MCAF transcription. Subsequently, HEKs were metabolically labelled with ^{35}S-methionine/cysteine prior to cytokine stimulation and the supernatants, at 18 h, immunoprecipitated either with a rabbit polyclonal antibody to a human MCAF fusion protein or preimmune serum. MCAF was precipitated as a doublet of M.W. 13 kD and 9 kD only in the IFN-gamma-treated HEK supernatants. Chemotactic activity of 10x concentrated supernatants was assayed via 48-well microchemotaxis plates with FMLP as the positive control and purified human peripheral blood monocytes as the target cell. Dilutional analysis revealed directed monocyte migration, induced by the IFN-gamma-treated HEK supernatant (80% maximal FMLP response), which was tenfold greater than IFN-gamma or HEK activity alone. Furthermore, monocyte chemotaxis was >80% inhibitable by anti-MCAF antiserum whereas little or no inhibition was observed by blocking studies utilizing anti-TGFβ. These studies demonstrate that HEKs elaborate biologically active MCAF, the modulation of which reveals cytokine specificity. In addition to its role in directing migration of monocytes towards the epidermis in inflammatory skin conditions, HEK-derived MCAF may also be responsible for the migration of dendritic Langerhans cells and dermal dendrocytes, both of which share many features with circulating monocytes, to their anatomical site in skin. Finally, a critical role for IFN-gamma in the pathogenesis of cutaneous inflammation is suggested by its regulation of both monocyte chemotaxis and adhesion molecules.

DIFFERENTIAL EFFECTS OF PROTEIN KINASE C INHIBITORS ON INTERLEUKIN 8 INDUCED EXOCYTOSIS IN HUMAN NEUTROPHILS

C Lam, L Klein and I Lindley. Sandoz Forschungsinstitut
Brunner Strasse 59, A-1235 Vienna, Austria.

Interleukin 8 (IL-8) is present in large amounts in psoriatic plaques and in synovial fluid from joints of arthritis patients. Locally produced IL-8 may be involved in recruiting and activating inflammatory cell infiltrate (T cells and neutrophils) to injure host tissue.

In this study, we investigated the effect of inhibitors of protein kinase C on stimulation of exocytosis in the neutrophils by IL-8. Purified neutrophils (10^6) were treated with serial concentrations of staurosporine (0.2-200 nM), calphostin C or calphostin 1 (0.1-10 μM) at 37°C for 10 min, and elastase secretion in response to IL-8 was measured. Whereas staurosporine alone was weakly active in stimulating the secretion of elastase at 200 nM, calphostin C and 1 (up to 10μM) were not secretagogues. Staurosporine-treated cells became primed for enhanced secretion of elastase in response to IL-8. Thus elastase secretion was enhanced dose-dependently to plateau at about 200-fold above the level released by cells stimulated with IL-8 alone. In contrast both calphostin C and 1 inhibited IL-8-induced exocytosis in neutrophils dose-dependently, with complete inhibition observed at 10 μM.

Taken together, the findings indicate that staurosporine is a secretagogue whereas calphostin C and 1 can inhibit components of the signal transduction involved in IL-8 mediated exocytosis. It is thus possible that analogs of the calphostin series could be exploited as novel anti-inflammatory agents for ameliorating neutrophil-mediated tissue injury provoked by IL-8.

INTERLEUKIN 8 STIMULATES AN INCREASE IN CYTOSOLIC FREE CALCIUM IN HaCat EPIDERMAL CELLS AND NEUTROPHILS

A Tuschil, C Lam, A Haslberger, and I Lindley. Sandoz Forsch. Institut, Brunner Strasse 59, A–1235 Vienna, Austria.

A number of locally produced cytokines have been implicated as chemoattractants responsible for the recruitment of inflammatory cells such as neutrophils and lymphocytes in psoriatic involved skin or in rheumatoid arthritis. Neutrophil activating peptide-1/interleukin 8 is present in significant amounts in psoriatic plaques and in synovial fluid from joints of arthritis patients. The putative pro-inflammatory cytokine has been shown to be a chemoattractant for both epidermal cells and fibroblasts in culture.

The aim of the present study was to establish whether epidermal cells, alo respond to IL-8 by raising their cytosolic free Ca^{2+} in a similar manner as human neutrophils. Quiescent human keratinocyte (HaCat cells) cultures grown to confluence on coverslips in low calcium medium (0.06 mM Ca^{2+}) were loaded with indo-1AM. Addition of IL-8 (0.06 – 47 nM) to the indo-1-loaded HaCat cells induced a rapid rise in cytosolic free Ca^{2+} from a basal level of 145±38 to a peak level of about 889±10 nM. The induced rise in Ca^{2+} was transient and concentration-dependent. Half maximal effect was observed at 1.2 nM. Under the same conditions, human neutrophils also responded to IL-8 in a similar dose-range by a rapid and transient mobilization of Ca^{2+}.

Taken together, the findings indicate that IL-8 has a wider range of responsive target cells than hitherto thought.

INTERLEUKIN 8 PRIMING OF HUMAN NEUTROPHILS FOR AN ENHANCED RESPIRATORY BURST: ROLE OF CYTOSOLIC FREE CALCIUM

C Lam, A Tuschil and I Lindley. Sandoz Forschungsinstitut, Brunner Strasse 59, A–1235 Vienna, Austria.

A number of inflammatory mediators prime human neutrophils for enhanced respiratory burst. Neutrophil activating peptide-1 (NAP-1) or interleukin-8 (IL-8) is one of these mediators which, by itself, activates respiratory burst only very weakly, but can enhance or prime the neutrophils for subsequent stimulating with a triggering agent (1). The intracellular and molecular mechanisms of priming are poorly understood. To examine the role of an increase in cytosolic free calcium ion (Ca^{2+}) as a basis for IL-8 priming in neutrophils, cells loaded with the sensitive Ca^{2+} chelator indo-1 or fura-2 were examined for calcium response on stimulation with the cytokine. IL-8 (up to 10 µM) alone did not directly stimulate reactive oxygen metabolite production in human neutrophils. However, it induced a transient rise in the cytosolic free Ca^{2+} from a resting value of 136 nM to a peak level of 1400 nM. The Ca^{2+} response was transient and dose-dependent. Associated with the rise in Ca^{2+}, the neutrophils became primed and produced significantly increased reactive oxygen metabolites in response to subsequent stimulation with FMLP (5 nM). Half-maximal priming occurred with IL-8 at 0.5 ng/ml and was nearly optimal at 10 ng/ml. Buffering of cytosolic free Ca^{2+} with indo-1 (2 µM) in the absence of extracellular Ca^{2+} blocked priming of the cells with IL-8. Taken together, the findings suggest that mobilization of cytosolic free calcium by IL-8 may be a central event in the molecular mechanism by which it primes human neutrophils.

1. Van Dervort, A.L. & R.L. Danner. 1990. Interleukin 8 (IL-8) primes human neutrophils for enhanced production of superoxide (O_2). Abs.: Natl.Meeting of Soc.Critical Care Medicine. USA.

MOLECULAR CLONING AND THE EXPRESSION OF MONOCYTE CHEMOTACTIC PROTEIN, MCP-1, IN ARTERIAL ENDOTHELIUM

JY Shyy, YS Li, DW Massop, JF Cornhill, and PE Kolattukudy
Dept. of Surgery, Biomedical Engineering, Biotechnology Center, The Ohio State University, Columbus, Ohio 43210, USA.

Adherence of peripheral blood monocytes to vascular endothelial cells is one of the earliest events that occur in atherogenesis. Monocyte chemoattractants released at the foci of endothelial injury are proposed to initiate and promote this type of cellular interaction.

In the current study, we report the molecular cloning of human aortic endothelial MCP. Stimulation of expression of the MCP gene by cytokines and the MCP expression *in vivo* in aortic endothelium were also observed. Cloned cDNA for the human aortic endothelial MCP contained a 5' noncoding region of 31 bp, a 297 bp open reading frame and a 3' noncoding region of 331 bp. The nucleotide sequence is identical to that of glioma cells except for the replacement of T at 105 of the glioma clone by a C in the endothelial sequence. Northern blot analysis revealed two species of hybridizing bands of RNA at 0.65 and 0.75 kb in human aortic, human umbilical vein and human pulmonary arterial endothelial cell cultures. The two hybridizing bands were confirmed by probing with cloned 0.65 kb cDNA for MCP. Primer extension analysis showed that the difference in size between the two transcripts is not due to a difference at the 5' noncoding region. The level of the two MCP transcripts increased in aortic endothelial cells incubated with 100 units/ml of recombinant IL-1α, IL-1β or 200 ng/ml of TNFα. MCP transcripts in cells treated with IL-2, IL-6 and INF-gamma were quiescent. MCP transcripts were also detected in "living" human aortic endothelial tissue. The layer of endothelium was removed from the descending aorta of a 48-year old male donor during organ transplantation surgery. Blot analysis with cDNA probe showed MCP gene transcripts in the isolated poly(A)$^+$ RNA. Hybridization of MCP was not detected in control RNA from an unrelated tissue or in poly(A)$^-$ RNA isolated from endothelial cell cultures.

MODULATION OF IL-8 GENERATION IN THE PRE-MONOCYTIC CELL LINE, U937

R Pleass and J Westwick. Dept. of Pharmacology, The Hunterian Institute, Royal College of Surgeons of England, Lincoln's Inn Fields, London WC2A 3PN, UK

U937 cells are a continuous line of human cells of committed monocytic origin. We have investigated the effects of agents which modulate GTP-binding proteins and protein kinase C (pkC) to explore the signal transduction mechanisms involved in LPS- and cytokine-stimulated IL-8 generation. A sensitive and specific ELISA developed by Ceska *et al* (1) was used to quantitate extracellular IL-8. Additions of LPS (0.1 – 100 ng/ml), IL-1α (0.1 – 10 ng/ml) and, to a lesser extent, TNFα (3 – 30 ng/ml) produced a dose-related generation of extracellular IL-8. Untransformed, DMSO-transformed and c-AMP-transformed U937 cells were all responsive to the above stimuli. Pre-treatment of U937 cells with pertussis toxin (100 ng/ml) for 4 h produced approximately 80% inhibition of sub-optimal LPS-, IL-1α- and TNFα-induced generation of IL-8. In contrast, cholera toxin (100 ng/ml) produced a 10 – 40% enhancement of IL-8 generation induced by LPS, IL-1α or TNFα. Additions of the phorbol ester (TPA, 0.1 – 30 nM), but not 4α phorbol, produced a marked stimulation of IL-8 formation. However, the addition of LPS or IL-1 did not modify TPA-induced IL-8 formation. Staurosporine (0.1 – 3 μM) the non-selective pkC inhibitor, produced a dose-related inhibition of basal, LPS- and TNFα-stimulated generation of IL-8. Low dose (0.1 – 0.3μM) staurosporine doubled IL-1-induced formation of IL-8, although 1 – 3 μM staurosporine produced significant inhibition.

These results indicate that GTP-binding proteins and PKC have a regulatory role in the generation of IL-8 derived from U937 cells.

1. Ceska, M. *et al.* 1989. *Cytokine* 1: 136.

SELECTIVE INHIBITION OF CYTOKINE GENERATION IN HUMAN CELLS BY AN IL-1 RECEPTOR ANTAGONIST

Z Brown, L Fairbanks, R Thomas, R Pleass and J Westwick.
Dept. of Pharmacology, The Hunterian Institute, Royal College of Surgeons of England, Lincoln's Inn Fields, London WC2A 3PN, UK.

Eisenberg *et al.* (1) have recently cloned and expressed a novel IL-1 receptor antagonist (IL-1ra). We have examined the specificity and potency of this novel recombinant peptide, on the IL-1α-, TNFα- and LPS-induced generation of IL-8, IL-6 and PGE_2 by human umbilical endothelial cells (HUVEC), mesangial cells (MC) and synovial fibroblasts (SF). The IL-1ra (0.3 – 10,000 ng/ml) was not an agonist in these preparations. Sub-optimal concentrations of IL-1α, TNFα and LPS, selected from dose response studies, were added to confluent cultures after a 15 min pre-treatment of the cells with vehicle or IL-1ra (0.3 – 10,000 ng/ml). Extracellular concentrations of IL-8, IL-6 and PGE_2 at 24 h were measured by ELISA (2), RIA and RIA respectively. Calculation of molar ratios of antagonist/agonist demonstrated that IL-1ra was an effective antagonist of IL-1α-, but not TNFα- or LPS-induced effects in HUVEC, MC and SF. Surprisingly, the potency of the IL-1ra was dependent upon the cell type. The mean molar ratio for a 50% inhibition for IL-1-induced effects in HUVEC, MC and SF was approximately 3, 15 and 100 respectively. These results indicate that the IL-1ra is particularly effective at suppressing the inflammatory mediators derived from IL-1-stimulated HUVEC and MC.

1. Eisenberg *et al.* 1990. *Nature* 343: 341–346
2. Ceska, M. *et al.* 1989. *Cytokine* 1: 136.

INVITED SPEAKERS

Professor M. Baggiolini
Director Theodor Kocher Institute
Universität Bern
CH-3000 Bern 9
Postfach 99
Freiestrasse 1
Switzerland

Dr Z Brown
Dept of Renal Medicine
Institute of Urology
University College
London

Dr R Camp
Institute of Dermatology
St Thomas's Hospital
Lambeth Palace Road
London SE1 7EH

Professor A Cerami
Laboratory of Medical Biochemistry
The Rockefeller University
1230 York Avenue
New York 10021-6399
USA

Professor TF Deuel
Division of Hematology/Oncology
The Jewish Hospital of St Louis
216 South Kingshighway Boulevard
PO Box 14109
St Louis
Missouri 63178
USA

Professor SL Kunkel
Department of Pathology
M5214/0602 Medical Science
The University of Michigan
1301 Catherine Road
Ann Arbor, Mi 48109-0602
USA

Professor EJ Leonard
Immunopathology Section
Laboratory of Immunobiology
NCI, FRCF, Bldg 560. Rm 12-71
Frederich, Maryland 21701
USA

Dr Ivan Lindley
Sandoz Forchung Institut
Brunnerstrasse 59
A1235 Wien
Austria

Dr K Matsushima
Lab. of Molecular Immuno-Regulation
Nat. Cancer Institute
Building 560, Room 31-19
Frederick
Maryland 21701-1013
USA

Professor Joost J Oppenheim
Laboratory of Molecular
Immunoregulation
National Cancer Institute
Bldg 560, Rm 21-89A
Frederick MD21701-1013
USA

Dr A Rot
Sandoz Forchung Institut
Brunnerstrasse 59
A1235 Wien
Austria

Professor R Sager
Division of Cancer Genetics
Dana Farber Cancer Institute
44 Binney Street
Boston MA 02115
USA

Dr J-M Schröder
Dept of Dermatology
University of Kiel
Schittenhelmstrasse 7
230 Kiel
Germany

Dr RM Strieter
Department of Pathology
M5214/0602 Medical Science
The University of Michigan
1301 Catherine Road
Ann Arbor, Mi 48109-0602
USA

Professor J van Damme
Rega Institute for Medical Research
University of Leuven
B-3000 Leuven
Belgium

Dr A Walz
Theodor-Kocher-Institute
Universität Bern
CH-3000 Bern 9
Postfach 99 Freiestrasse 1
Switzerland

Dr J Westwick
Department of Pharmacology
Hunterian Institute
Royal College of Surgeons
35 Lincolns Inn Fields
London WC2A 3PN

Dr T Yoshimura
Immunopathology Section
Laboratory of Immunobiology
NCI, FRCF, Bldg 560, Rm 12-71
Frederich, Maryland 21701
USA

INDEX